Home Media Networks

2nd Edition

Lawrence Harte

Althos Publishing
106 W. Vance Street
Fuquay-Varina, NC 27526 USA
Telephone: 1-919-557-2260
Fax: 1-919-557-2261
email: info@althos.com
web: www.Althos.com

Althos

All rights reserved. No part of this book may be reproduced or transmitted in any form or by any means, electronic or mechanical, including photocopying recording or by any information storage and retrieval system without written permission from the authors and publisher, except for the inclusion of brief quotations in a review.

Copyright (c) 2009, 2010, 2011 By Althos Publishing

Printed and Bound by Lightning Source, TN.

> Every effort has been made to make this manual as complete and as accurate as possible. However, there may be mistakes both typographical and in content. Therefore, this text should be used only as a general guide and not as the ultimate source of information. Furthermore, this manual contains information on telecommunications accurate only up to the printing date. The purpose of this manual to educate. The authors and Althos Publishing shall have neither liability nor responsibility to any person or entity with respect to any loss or damage caused, or alleged to be caused, directly or indirectly by the information contained in this book.

International Standard Book Number: 1-932813-64-0

About the Author

Communication technology and business expert Lawrence Harte has designed, setup, and tested communication networks using a mix of data cables, coax, phone lines, wireless, and optical transmission technologies. Mr. Harte is the editor and publisher of IPTV Magazine, Mobile Video Magazine, and author of over 112 books on communication technologies (more than 20 on TV systems and technologies). Mr. Harte has interviewed over 3100 companies that produce TV products and services. He has worked for Ericsson/General Electric, Audiovox/Toshiba and Westinghouse and consulted for hundreds of other companies. He has earned many degrees and certificates including an Executive MBA from Wake Forest University and a BSET from the University of the State of New York. Mr. Harte can be contacted at LHarte@Althos.com.

Home Media Networks

Acknowledgements

Many smart people have helped us to create this book. Several of them gave substantial amounts of time to answer questions and share their hard to find experience with us. Some of these people include Rob Gelphman with MoCA, Multimedia over Coax Alliance, Anton Monk from Entropic, Lawrence Thorne at Firecomms, Andy Tarczon with TDG Research, Rich Nesin at Nesin Consulting, David Callisch from Ruckus Wireless, and Michael Weissman with Sigma Designs.

Some of the key people at Althos publishing who helped create this book include Michele Chandler (project management), Carolyn Luck (editor), Geovanny Solera (diagrams), and Jon-Luke Ramos (research), and Vivian McCarter (layout).

Table of Contents

ABOUT THE AUTHOR..III

ACKNOWLEDGEMENTS.....................................V

CHAPTER 1 - HOME MEDIA NETWORKS..................1

HOME MULTIMEDIA SERVICE NEEDS.............................2
HOME TELEPHONE SERVICE......................................4
   Telephone Data Transmission Rates......................4
   Telephony Transmission Characteristics................4
   Telephone Service Daily Data Consumption............4
INTERNET ACCESS...5
   Internet Access Data Transmission Rates..............5
   Internet Access Transmission Characteristics.........6
   Internet Access Daily Data Consumption...............6
HOME TELEVISION SERVICE.......................................7
   Television Data Transmission Rates.....................7
   Television Transmission Characteristics................8
   Television Service Daily Data Consumption...........8
HOME INTERACTIVE VIDEO..9
   Interactive Video Data Transmission Rates............9
   Interactive Video Transmission Characteristics.......10
   Interactive Video Service Daily Data Consumption..10

  HOME MEDIA STREAMING 11
   Media Streaming Data Transmission Rates *11*
   Media Streaming Transmission Characteristics *11*
   Media Streaming Daily Data Consumption *12*

CHAPTER 2 - HOME NETWORK SYSTEM NEEDS 13

  CO-EXISTENCE 13
   Transmission Types *13*
   Intersystem Interference *14*
   Protocols *14*
   Alien Networks *15*
  NO NEW WIRES (NNW) 16
   Site Review *16*
   Hybrid Home Media Networks *17*
  DATA RATES 18
   Gross Data Rate (Gross Rate) *18*
   Net Data Rate (Net Rate) *18*
   Average Data Rates *18*
   Peak Data Rates *19*
  QUALITY OF SERVICE (QOS) 19
   Error Rates *19*
   Data Throughput *19*
   Transmission Delay *20*
   Jitter *20*
  HOME COVERAGE 21
   Connection Types and Locations *21*
   Home Coverage Area *21*
   Visibility *21*
  SECURITY 22
   Privacy *22*
   Rights Management *22*
   Content Protection *22*

INSTALLATION 23
   Installation Skills *23*
   Configuration *23*
   Remediation *24*
   Upgrading *24*
GATEWAYS 24
   Firewalls *25*
   Protocol Adaptation *25*
   Remote Diagnostics *25*
COST 26
   Equipment Costs *26*
   Installation Costs *26*
   License Costs *26*
   Support Costs *27*

HOME NETWORKING TECHNOLOGIES **29**
ADAPTIVE MODULATION 29
   Modulation Type *30*
ECHO CONTROL 31
   Signal Reflections *31*
   Echo Cancelling *31*
SYNCHRONIZED TRANSMISSION 32
   Type of Service (ToS) *32*
   Master Controller *32*
   Contention Free Transmission *33*
   Contention Based Systems *33*
INTERFERENCE AVOIDANCE 34
   Interference Detection *34*
   Channel Selection *35*
   Network Coordination *35*

POWER LEVEL CONTROL 35
   Signal Level Detection *35*
   Transmitter Power Control *36*
CHANNEL BONDING 36
   Channel Multiplexing *36*
   Channel Combining (Demultiplexing) *36*
TOPOLOGY DISCOVERY 37
   Link Detection *38*
   Node Capabilities *38*
   Network Mapping *38*

HOME NETWORK TRANSMISSION TYPES 39

WIRED LAN 39
WIRELESS 40
   Frequency Bands *42*
   Data Transfer Rates *42*
   Multi-beam Smart Antenna System *43*
ELECTRICAL POWER LINE CARRIER (PLC) 45
   Electrical Noise *45*
   Phase Coupling *45*
   Changing Topology *46*
COAXIAL 47
   Tree Distribution *48*
   Cable Television Co-Existence *48*
   Coaxial Wire Types *48*
TELEPHONE LINE 50
   Wire Types *50*
   Interfering Signals *51*
   Splices *51*
   Signal Leakage *51*
HOME OPTICAL NETWORKS 53
   Fiber Type *53*
   Optical Type *53*
   Connection Type *54*

HOME NETWORK SYSTEMS 55

HOMEPLUG™ 55
HomePlug 1.0 55
HomePlug Audio Visual (HomePlug AV) 58
DIGITAL HOME STANDARD (DHS) 62
HD-PLC 63
HOMEPNA™ 64
HomePNA 1.0 65
HomePNA 2.0 65
HomePNA 3.0 66
HomePNA 3.1 67
MULTIMEDIA OVER COAX ALLIANCE (MoCA)™ 70
802.11 WIRELESS LAN 72
Quality Performance Monitoring 73
802.11E QUALITY OF SERVICE 74
Enhanced Distributed Channel Access (ECCA) 75
802.11N MULTIPLE INPUT MULTIPLE OUTPUT (MIMO) 75
HOMEGRID 76
Unified Network 77
Home Media Domains 77
Foreign Networks 77
FIBERCOMMS HOME OPTICAL NETWORK 78
LED Optical 79
Plastic Fiber 79
Opto-Lock Connectors 80

HOME MEDIA MANAGEMENT 83

DIGITAL LIVING NETWORK ALLIANCE (DLNA) 83
Digital Media Server (DMS) 84
Digital Media Controller (DMC) 84
Digital Media Player (DMP) 84
Digital Media Renderer (DMR) 84
Digital Media Printer (DMPr) 85

DLNA LAYERS 86
   Media Format Layer *86*
   Media Management Layer *86*
   Device Discovery and Control Layer *86*
   Media Transport Layer *86*
   Network Stack Layer *87*
   Connectivity Layer *87*
UNIVERSAL PLUG AND PLAY (UPNP) 88
   Automatic Addressing *89*
   Device Discovery *89*
   Capabilities Description *89*
   Control Functions *89*
   Event Notifications *90*
   Presentation *90*
UPNP AUDIO VISUAL (UPNP AV) 91
   Media Server *91*
   Control Point *91*
   Media Renderer *92*
   Rendering Control *92*
   Quality of Service (QoS) Policy *92*
   Remote User Interface (RUI) *92*

DIGITAL RIGHTS MANAGEMENT (DRM) **93**
  MEDIA PORTABILITY 94
  AUTHENTICATION 95
  DIGITAL CERTIFICATE 98
  DIGITAL SIGNATURE 99
  ENCRYPTION 100
   Encryption Keys *101*
   Symmetric and Asymmetric Encryption *102*
   Public Key Encryption *103*
  USAGE RESTRICTIONS 103
   Copy Control Information (CCI) *104*

DIGITAL TRANSMISSION CONTENT PROTECTION (DTCP) 105
   Authentication and Key Exchange (AKE) *105*
   Device Certificates *106*
   Content Protection (Encryption) *106*
   System Renewability *106*
   Digital Transmission Control Protection Plus (DTCP+) *108*

APPENDIX 1 - ACRONYMS 111

INDEX. 115

Chapter 1

Home Media Networks

Home multimedia networks (HMNs) are the combination of equipment and software used to transfer data and other media within a customer's facility, home or personal area. A home multimedia network is used for media devices such as digital televisions, computers and audio visual equipment to transfer media to one another, and to other networks, through the use of wide area network connections. HMN systems may use wired Ethernet, wireless LAN, electrical power lines, telephone lines, coaxial cables or optical connections.

HMN systems have transitioned from low speed data, simple command and control systems to high-speed multimedia networks. They have evolved to include the ability to transfer a variety of media types (data, telephone audio, and video) that have different transmission and management requirements. Each of the applications that operate through an HMN can have different communication requirements which typically include a maximum data transmission rate, continuous or bursty transmission, packet delay, jitter and error tolerance. The HMN system may manage these connections using a mix of protocols that can define and manage quality of service (QoS).

Figure 1.1 shows the common types of HMNs that can be used for IP television systems. This diagram shows that an IP television signal arrives at the premises at a broadband modem. The broadband modem is connected to a router that can distribute the media signals to forward data packets to different devices within the home such as IP televisions. This example shows that routers may be able to forward packets through power lines, telephone lines, coaxial lines, data cables or even via wireless signals to adapters that receive the packets and recreate them for use by IP televisions.

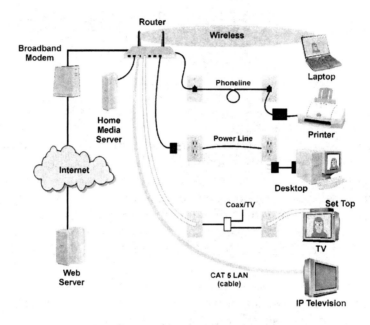

Figure 1.1, Home Media Distribution Systems

Home Multimedia Service Needs

Multimedia networking is the process of transferring multiple forms of media (such as digital audio, data and digital video) through a communication network, such as a home network. Multimedia networking is used to connect computers, media players and other media sources to each other, and to send and receive media from wide area network connections.

The communication requirements for multimedia networks in the home are based on the types and amounts of applications used, as well as the times during which the applications and services will be used. The typical types of applications used in the home include telephone, Internet access, television, interactive video and media streaming.

Figure 1.2 shows some of the types of communication devices used in a home and their estimated data transmission requirements. This table shows that some devices may require connections through a gateway to other networks, such as the Internet or a television system. This table also shows that the highest consumption of bandwidth occurs from television channels, especially when simultaneous HDTV channels are access on multiple television sets. This figure suggests that a total home media capacity of 70 Mbps to 100 Mbps is required to simultaneously provide for all devices within a home, and a residential gateway must have broadband capability of 50+ Mbps.

Service	Bandwidth	Number of Devices	Bandwidth Residential Gateway
TV	2 to 20 Mbps	3	2 to 54 Mbps
Digital Video Recorder (DVR)	2 to 20 Mbps	1	0
Home Theater	1 to 6 Mbps * Audio	1 System	0
Internet Browsing	1 to 2 Mbps	1 to 5	1 to 10 Mbps
Printer	0.5 to 1 Mbps	1 to 5	0
Digital Imaging	1 to 20 Mbps	1 to 3	0
Digital Telephone	0.2 Mbps	1 to 5	0.2 to 1 Mbps
Online Gaming	0.2 to 1 Mbps	1-3	0.2 to 3 Mbps
Video Capturing	0.1 to 1 Mbps	1-10 * Security Cameras	0
Portable Audio	0.1 to 20 Mbps	1 to 3	0
Total	70 Mbps to 100 Mbps		2 to 60 Mbps+

Figure 1.2, Home Media Services Bandwidth Requirements

Home Telephone Service

Home telephone service is the provision of publicly available audio communication service (telephony) to telephone devices such as telephones and fax machines. Home telephone service may involve the transmission of speech (telephone) or data (fax, computer modems). The typical data transmission rate for home telephone audio signals is low (90 kbps). The connection time is moderate (several minutes), and the daily consumption is relatively small (approximately 40 to 80 MBytes per day) when compared to other home communication services.

Telephone Data Transmission Rates

Uncompressed digital telephone signals have data transmission rates of approximately 90 kbps in each direction. This includes the 64 kbps audio signal plus the signaling overhead (packet addressing and control messages).

Telephony Transmission Characteristics

The transmission characteristics required for home telephone services include low delay packets (approximately 100 msec or less), and some transmission errors are acceptable (voice can be slightly distorted).

Telephone Service Daily Data Consumption

Because home telephone usage is typically totals less than 1 hour and the call durations only last for a few minutes, telephony services only require a small amount of the daily data transfer capacity of HMNs. The total daily bandwidth consumption is approximately 40 to 80 MBytes per day.

Figure 1.3 shows that uncompressed home telephone service requires approximately 90 kbps of data transmission for each direction of communication. This is divided into 64 kbps for digital audio and 26 kbps for control (addressing and signaling).

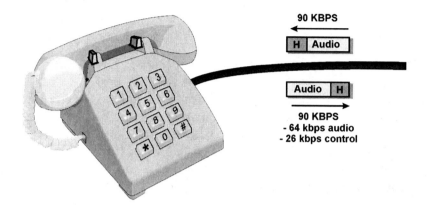

Figure 1.3, Home Telephone Service Needs

Internet Access

Internet access is the ability of a user or device to connect and access media and services through the Internet. Home Internet access typically involves short high speed data transfers. Errors that may occur, such as lost packets, can be corrected through data packet retransmission. The amount of daily data consumption for Internet access is relatively low (approximately 100 MBytes per day).

Internet Access Data Transmission Rates

Internet communication tends to require high-speed data transmission in short bursts. Short delays are acceptable and transmission errors are usually automatically overcome by Internet protocols. In general, the amount of data transmission required for Internet connections continues to increase as rich media sources (video streaming) increase. The data transmission rate for downlink (from the Internet) tends to be faster than data uplink (to the Internet) connections. This is because people tend to gather more information from the Internet than they send to the Internet.

Internet Access Transmission Characteristics

The transmission characteristics required for home Internet service include temporary high-speed data transfers. Some transmission errors (such as corrupted or lost packets) are acceptable because Internet access programs will automatically request retransmissions of the lost data.

Internet Access Daily Data Consumption

Home Internet usage consists of approximately 1 hour per day and the connection time (session) can last from minutes to hours. During the connection, the data transfer rate fluctuates as Internet users load new web pages and transfer media files. As a result, the amount of data transmission can be 10% of the connection speed (100kps to 200 kbps average). The total daily bandwidth consumption for home Internet access in 2010 was approximately 50 to 100 MBytes per day.

Figure 1.4 shows that Internet access service needs for home Internet access involve the provision of high-speed data connections that transfer data in

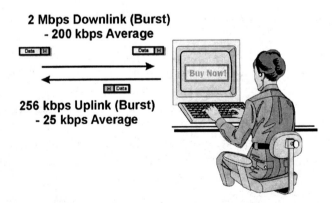

Figure 1.4, Home Internet Access Service Needs

bursts. This example shows a user that is receiving data from the Internet at a maximum rate of 2 Mbps, and sending data to the Internet at 256 kbps. This example also shows that this user has an average data connection rate of 10% of maximum (approximately 200 kbps).

Home Television Service

Home television service is the transmission of television programming (typically video combined with audio) to one or more TV viewing devices in the home. A single home can have multiple televisions which require multiple simultaneous data streams. High definition television channels are becoming more available which is likely to result in an increase in the bandwidth required for television signals. Household television viewing often occurs at the same time, which is usually during the evening hours.

Digital television signals require high-speed data transmission rates (2-20 Mbps) that are used for several hours per day. For some types of digital compression such as MPEG-2, even small amounts of transmission errors (0.1%) can result in television signals that are unwatchable.

Because television services can consume high percentages of HMN bandwidth, it is important that the HMN be able to categorize and prioritize television services. Many HMN networks have evolved to include an audio/visual (A/V) version which can efficiently and reliably distribute television signals over data networks in the home.

Television Data Transmission Rates

TV service requires continuous high-speed data transmission (2 Mbps) over relatively long periods of time (several hours). In general, the amount of data transmission required for TV service continues to increase as more high definition programming becomes available. High definition TV service data transmission rates range from approximately 6 Mbps (MPEG-4) to 20 Mbps (MPEG-2).

During certain time periods, all viewers in the home may be simultaneously watching TV programs. Because each TV viewing experience is a separate streaming channel, the peak data transfer rate through the home gateway can range from 6 Mbps to over 20 Mbps.

Television Transmission Characteristics

The transmission characteristics required for home TV service include variable rate high-speed data connections, low jitter rates and low error rates. The data transfer rates for TV programs can also temporarily increase with the occurrence of rapidly changing video activity, such as high action scenes.

Some transmission errors (such as corrupted or lost packets) are acceptable for TV transmission as minor packet losses will typically result in small video distortions. When control signal packets, such as channel change requests, are lost, TV application programs will automatically request retransmission of the lost data. The data transmission rate for TV service is almost exclusively in the downlink direction (from the TV broadcaster).

Television Service Daily Data Consumption

Home TV usage is approximately 4 to 5 hours per day and the connection time (session) can last from minutes to several hours. During the connection, the data transfer rate averages approximately 2 Mbps (or higher) for standard definition TV. This results in a total daily bandwidth consumption rate per home TV of 9 GBytes per day.

Figure 1.5 shows the data transfer rate needed to provide for multiple IPTV users in a single building. This diagram shows 3 IP televisions that require 1.8 Mbps to 3.8 Mbps to receive an IP television channel. This means that the broadband modem must be capable of providing 5.4 Mbps to 11.4 Mbps to allow up to 3 IP televisions to operate in the same home or building.

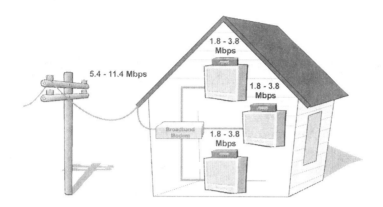

Figure 1.5, Home Television Service Need

Home Interactive Video

Home interactive video services are processes that allow users to interactively provide and receive audio and/or visual information. Some of the key interactive video services include Internet video calls and network gaming services.

Interactive Video Data Transmission Rates

Video call data transmission rates, such as those pertaining to Skype video calls, are approximately 50 kbps to 200 kbps, but these rates are increasing with the availability of more bandwidth. Interactive video and graphics are also used by gaming devices and their data transmission rates can vary. Network games that have large amounts of video and animation can consume 100 kbps.

Home Media Networks

Interactive Video Transmission Characteristics

The transmission characteristics required for interactive video services include continuous and variable rate medium-speed data connections that have low jitter rates and low error rates. The data transfer rates for interactive programs can also temporarily increase when users send or receive media objects such as images or media clips.

Interactive Video Service Daily Data Consumption

Interactive video calls can last for 10 to 20 minutes per day. However, network gaming sessions can last for 1 to 2 hours per day [1].

Figure 1.6 shows that some of the sources for interactive video service include video calls and gaming. Video calls use data rates that can range from 100 kbps to over 700 kbps. Gaming connections generally have relatively low data rates for control signals but have bursts of high data rates for image transfers.

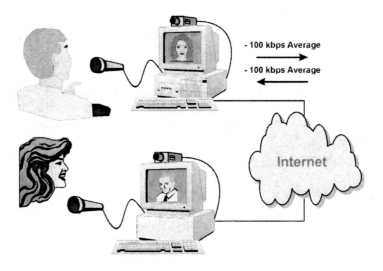

Figure 1.6, Home Interactive Video Service Needs

Home Media Streaming

Home media streaming is the continuous transfer of media (video or audio) between devices in the home. Home media applications include security camera connections or music distribution (home theater) systems. Although the amount of data transmission for these may be low, they may be used for long durations.

The data transmission requirements for home media networks may be increased as a result of multiple connection paths (double or triple hops) or playback modes that can require multiple streams.

A double hop is a transmission path that is routed through a communication network that requires two connection paths to reach its destination rather than a single direct path. Double hop connections through a home media network can use up to two times as much data transmission capacity.

Media Streaming Data Transmission Rates

Home media streaming requires continuous medium to high speed data transmission (100 kbps to 2 Mbps) over relatively long periods of time (several hours). In general, the amount of data transmission required for home media streaming is likely to increase as more content and higher quality versions (HD) become available for distribution within the home.

Media Streaming Transmission Characteristics

The transmission characteristics required for home media service include variable rate data connections, low jitter rates and low error rates. The data transfer rates for video programs can also temporarily increase with the occurrence of rapidly change video activity (high action scenes). Some transmission errors (such as corrupted or lost packets) are acceptable for home media transmission as minor packet losses will typically result in small video distortions.

Media Streaming Daily Data Consumption

The amount of media streaming data consumption through the gateway is 0 because the media streaming takes place between devices in the home.

Figure 1.7 shows how homes may use streaming services between devices. This example shows that two televisions are watching a homemade video program from a digital video recorder (media server). This media server has established streams with two devices in the home and each stream is 700 kbps. Even though they are the same media program, the media server is providing a separate stream to each viewing device, which results in the home media network transferring 1.4 Mbps. Because these streams are only transferred in the home, the amount of data sent through the home media gateway is 0.

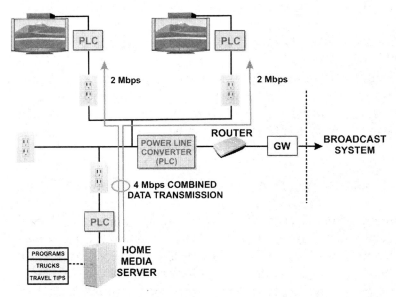

Figure 1.7, Home Media Streaming Service Needs

References:

[1]. "Playing Games with Media Usage" November 4, 2009, eMarketer, http://www.emarketer.com/Article.aspx?R=1007362.

Chapter 2

Home Network System Needs

The requirements for home network systems vary depending on the needs of the viewers, the needs of the TV and the Internet service provider, and the availability of existing home network systems and connection options. Some of the general factors that are considered when selecting a home multimedia networking system include co-existence, lack of new wiring requirements, data rates, quality of service (QoS), security, cost and installation.

Co-Existence

Co-existence is the ability of a device or system to operate near or with another device and/or system. HMNs typically co-exist with one or more types of systems including wireless LAN, CATV, satellite, home audio/visual and others. The HMN system may need to share the same connection paths, which may include the cables and/or radio waves.

Transmission Types

To co-exist, these systems may use a mix of different physical transfer paths, including wireless and cable, or segmentation of a single medium (a phone line or coaxial line) where a portion of the medium (a frequency range) is used by one system and another segment (a different frequency range) is used by the HMN system.

Intersystem Interference

Intersystem interference is the interaction of signals or data in one system with signals or data in another system, resulting in performance degradation or unwanted changes in the operation of either system. HMN systems may experience or cause interference with other systems that may be adjacently located, such as a home network that is installed in a neighbor's home.

Intersystem interference may occur between systems that are physically connected or it may occur by signal leakage. When interference occurs, it may result in distortion or reduced transmission capacity. In some cases, interference reduction may be accomplished by the inserting point of entry (POE) blocking or notching signal filters between systems. A POE filter is a device (analog or digital) that is designed to allow or block specific signals from entering or leaving a building or facility (at the point of entry).

In some cases, interference may be reduced by eliminating some of the channels (frequency bands) that are used within the system. This may be accomplished by installing a frequency band notching filter. An FBN filter may be used to minimize interference from or to subchannels within a multifrequency carrier.

Protocols

Protocols are the languages, processes and procedures that perform the functions used to send control messages and coordinate the transfer of data. Protocols define the format, timing, sequence and error checking used on a network or computing system.

Home networks may use a combination of protocol languages to coordinate their operation. For most HMN systems, the underlying processes are commonly based on Ethernet and Internet protocols (IP). Protocols for distribution services (such as RTP) simply ride on top of (encapsulated within) the data packets that travel through the data network.

Protocols tend to evolve over time with the addition of new commands and processes that can be performed. Older equipment that does not have the latest version of the protocol installed will commonly ignore commands that are not understood. This is especially important in home media network systems that mix data communications and audio visual communications. The coordination of audio visual signals commonly involves scheduled data transmission priorities which were not necessary in older home media network systems.

Some home media network protocols include the capability to automatically detect, communicate and coordinate with other networks. This can allow the networks to identify interference and select channels and transmission times that reduce the interference between each network.

Alien Networks

Alien networks are communication systems that cannot be identified or with which communication cannot be established. Alien networks operate independently of home media networks. Because a home media network cannot coordinate with an alien network, it may interfere with the HMN, and the HMN may interfere with it. An example of an alien network is an unidentifiable cordless telephone system that operates on the same frequency as a wireless LAN system. The only options for the wireless LAN system are to either suffer the interference (some packets will need to be retransmitted, slowing the data transmission rate) or to identify which frequencies are being interfered with, and use other frequency channels to avoid such interference.

Figure 2.1 shows how multiple home media network systems can co-exist with each other. This example shows a home that uses a mixture of phone line, power line and coax transmission types. Of these systems, the HomePNA version has multiple versions (HomePNA 2.0 – data and HomePNA 3.1 – audio visual). Advanced protocols are used, which are ignored by older versions. These systems experience interference from each other and from other networks (alien networks).

Home Media Networks

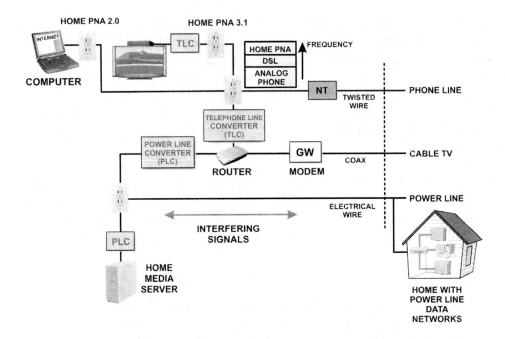

Figure 2.1, Home Media Network Co-Existence

No New Wires (NNW)

No New Wires is a concept that can be used when installing additional communication capabilities within a home or building, without adding any new wires. The NNW concept does allow for the use of existing wires as part of the new communication system.

Site Review

Site review is an inspection of a location to determine existing and/or upgradeable capabilities of devices, systems or facilities. A site review can determine the existing communication lines and equipment that can be used for HMNs. A site review may include identifying the types of devices (TVs, computers, game consoles), communication lines (telephone lines, coaxial lines, data lines), connecting line interconnection points (topology), and existing equipment (routers, splitters, signal boosters).

Chapter 2

Hybrid Home Media Networks

The concept of no new wires can be extended to allow HMN systems to use a mix of different physical transfer systems. For example, an HMN system might use a combination of phone lines, power lines and coaxial lines to interconnect devices throughout a home or building.

Figure 2.2 shows how TV broadcasters and consumers prefer to have no new wires when connecting voice, data and video devices. The first step is a site review that identifies the available connection types and required services. This home has a mixture of telephone lines, electric power lines and coaxial lines. Since no single connection type exists for all the connected devices, a mixture of connection types can be used.

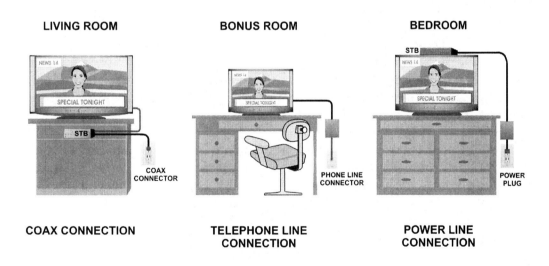

Figure 2.2, Home Media Existing Wiring Options

Data Rates

Data rate is the amount of information that is transferred over a transmission medium during a specific period of time. The data rates available on home media networks may be described in gross data rates or net data rates. The required data transmission rates for home media networks services can be characterized by gross data rate or net data rate, and by average and peak data transmission capabilities and requirements.

Gross Data Rate (Gross Rate)

Gross rate is the rate of all information that is sent on a transmission channel. Gross rate includes all data that is used for control signaling and the user data portion of the transmission. For example, the gross data transmission rate on an 802.11 WLAN system may be 11 Mbps or 54 Mbps.

Net Data Rate (Net Rate)

Net rate is the portion of information that is sent on a transmission channel or system that is available to users or devices. The net data transmission rate is the gross data transmission rate reduced by the control (signaling) data, and the data or transmission time that is used for non user data transmission purposes.

For home media networks, the net data transmission rate can be affected by external environments such as unexpected interference from other systems, or from changes in the existing systems (such as connection lines or devices that may be added or removed from the system).

Average Data Rates

Average data rate is the data transmission usage that occurs over a period of time. Average data transmission rates on home media networks are being reduced by the use of improved media compression capability, including the transition from MPEG-2 to MPEG-4. Average rates are being increased by the addition of new applications and services, such as video telephone calls.

Peak Data Rates

Peak data rate is the maximum amount of information (data) that can be transferred over a transmission medium during a short period of time. The required data transmission rates in networks can peak when multiple users simultaneously use services that can have variable data transmission rates, such as a burst of data that occurs during a new media source selection.

Quality of Service (QoS)

Quality of service is one or more measurements of desired performance and priorities of a communications system. Some of the important QoS characteristics for home media network systems include error rates, transmission delay and jitter.

Error Rates

Error rate is a ratio between the amount of information that is received in error as compared to the total amount of information that is received over a period of time. Error rate may be expressed in the number of bits that are received in error, or the number of blocks of data (packets) that are lost over a period of time.

Some digital video formats (such as MPEG-2) are more sensitive to packet error rates (burst errors) than packet errors. Acceptable error rates for video distribution are a packet error rate (PER) of 10^{-6} and a bit error rate (BER) of 10^{-9}.

Data Throughput

Data throughput is the amount of user data information that can be transferred through a communication channel or a point within a communication system. The net data throughput is lower than the gross data transmission rate because of the need to use some of the transmission data for control purposes (packet addressing), and because of interference or coordination with other transmission signals.

Transmission Delay

Transmission delay is the amount of time that is required for the transmission of a signal or packet of data from the point at which it enters a transmission system (transmission line or network) to the point at which it exits the system. Common causes of transmission delay include transmission time through a transmission line (less than the speed of light), channel coding delays, switching delays, queuing delays waiting for available transmission channel time slots, and channel decoding delays. Since an HMN system is generally small with a limited number of switching points, the overall transmission delay is not usually a significant issue.

Jitter

Jitter is a small, rapid variation in the arrival time of a packet of data that is typically the result of fluctuations in packet switching changes or delays. Digital video and audio is generally more sensitive to jitter than it is to overall transmission delay. Jitter can cause a packet to be received out of sequence and miss its presentation window.

The Ethernet data system used by many HMNs uses a random packet access control process that can introduce a substantial amount of jitter. This can dramatically affect the QoS of real time media signals such as telephony and video.

Home Coverage

Home coverage is the area (for wireless) or outlet locations (for wired systems) where users can connect to the home media network. These connection points should have a signal strength level that is sufficient to transmit and receive information, media or data to specific areas or connection points within the home.

Connection Types and Locations

There may be several connection types (such as wireless and wired) and locations within a home. The connection type and locations should be close to the media device, which may be a television, computer or media player.

Home Coverage Area

Home coverage area is the amount or percentage of physical area within the home that has sufficient signal strength level (powerline, radio) to allow for transmission of the desired services. The coverage area may vary depending on the amount of bandwidth required, as increased distances may reduce the bandwidth capacity of the connection.

Visibility

Visibility is the ability of a user or device to find and communicate with a system or other device that is connected to the home media network. Incomplete visibility is the inability of some devices within a network to identify or communicate with other devices within the network. Invisibility may be caused by the degradation of signal quality as it travels between devices or the configuration and operation of other devices that are attached to the network.

Early HMN systems used relatively low frequencies which had difficulties reaching many locations in the home. This limited the coverage area, reducing number of available HMN access points. Improved HMN systems that use higher frequencies and advanced signal processing capabilities have dramatically improved the amount of home coverage.

Security

Security is the ability of a person, system or service to maintain its desired well being or operation without damage, theft or compromise of its resources from unwanted people or events. Security for IPTV home multimedia distribution systems includes privacy, rights management and content protection.

Privacy

Privacy is the protection of user information, which can include personal contact information, media choices and usage information. Home media networks may use personal information to help with user to navigation and control of home networks. To protect user information, home media networks may use anonymized information to ensure that no specific customer is identified with the collected usage information.

Rights Management

Rights management is a process of organization, access control and assignment of authorized uses (rights) of content. Rights management may involve the control of physical access to information, identity validation (authentication), service authorization and media protection (encryption). Rights management systems are typically incorporated or integrated with other systems, such as content management systems, billing systems and royalty management systems.

Content Protection

Content protection is the end-to-end system that prevents content from being pirated or tampered with in a communication network (such as in a television system). Content protection involves uniquely identifying content, assigning the usage rights, scrambling and encrypting the digital assets prior to play-out or storage, both in the network and the end user devices, and delivering the accompanying rights to allow legal users to access the content.

Some security methods may be included in a home media system. User information may be kept private (anonymized) so it cannot be shared or used by companies or other users. Rights management identifies content and associates it with usage authorizations, such as number of views and ability to copy. Content protection is the conversion of information (encryption) into a format that cannot be used by people or devices.

Installation

Installation is the locating and configuration of equipment and/or wiring, either inside or outside of buildings or facilities. Installation requirements may include installer skill levels, configuration processes, the ability to solve (remediate) unexpected issues and the upgrading of devices and systems.

Installation Skills

While it is possible for customers to self-install some types of HMNs (plug and play), professional installation may be used to ensure that equipment has passed operational and performance tests. Necessary skills may include installing cables, connecting test equipment and understanding test results.

Configuration

Some HMN systems may be capable of self discovery and automatic configuration. Self discovery is the processes used to request and receive information that is necessary for a device to begin operating. Self discovery may involve finding devices within a system that can provide the identification address, name and services of another device. Self configuration is the selection of options (parameters) that enable the device to operate effectively, which might include the identification of channel frequencies to use and avoid.

Remediation

Remediation is the correction of a condition or problem. For HMN systems, remediation may involve the dispatching of a qualified technician or field service representative to validate, configure or correct faulty installations. Remediation may also involve the use of remote diagnostics to allow a service provider (such as a TV broadcaster) to view the devices that are operating within a home (connecting through a gateway such as a cable or DSL modem).

Upgrading

Upgradability is the capability of devices or systems to be modified, changed or use newer components as technology innovations become available. HMN systems may be upgradeable through the use of automatic software downloads through the Internet.

Required upgrade installation skills can range to include general consumer knowledge or proper cable and equipment training. Configuration can range from automatic detection and configuration to manual entry of option information. Remediation can range from remote diagnostics to deploying a qualified technician who can analyze and repair systems. Upgrading can range from automatic firmware downloads to the replacement of devices.

Gateways

A gateway is a communications device or assembly that transforms data that is received from one network into a format that can be used by a different network. Gateways can provide several functions for home media networks including firewalls, protocol adaptation and enabling of remote diagnostics.

Firewalls

A firewall is a data filtering device that is installed between a computer server or data communication device and a public network such as the Internet. A firewall continuously looks for data patterns that indicate the presence of unauthorized use or unwanted communications with the server.

Home media network gateways may include firewalls that are aware of applications. They should allow certain types of services (such as streaming video) from the TV broadcaster without the addition of undue packet delays due to firewall packet analysis and filtering.

Protocol Adaptation

Protocol adaptation is the process of converting the commands or processes from one protocol to another protocol. This may involve syntax changes (text format and command name changes), timing relationships and other functional processes. Gateways may adapt communication sessions that are managed using one protocol into another protocol.

Remote Diagnostics

Remote diagnostics is a process and/or program that is built into a device or network component that allows users or devices from remote locations to access information or control operations. Remote diagnostics can be used to discover the functionality or performance of a device or system connection.

Gateways may be designed to allow for the gathering, storage and transmission of home network configuration and performance information. The gateway may allow the TV broadcaster network to see inside the home network and setup and control devices.

Cost

Cost is the fees or resources that are required to produce or provide a product or service. HMN costs can include equipment and installation fees, license and franchise fees, and transaction and service support costs.

Equipment Costs

Equipment costs are the fees that are paid by the customer and/or broadcaster for equipment that will be installed in the home. Some home media network equipment (such as gateways or routers) may be provided to the consumer by the TV broadcaster for free or at reduced cost.

Consumers may purchase their own home media network equipment such as powerline data links. The cost of these devices can strongly influence the consumer's choice to use them. Wireless home media network equipment is available at a relatively low cost due to the large volume of devices purchased by consumers (economies of scale). Other types of home media network adapters (such as powerline or telephone line adapters) tend to be more expensive. Lower equipment costs may lead consumers to choose devices that may not be compatible with their systems. For example, consumers might purchase an 802.11g wireless router instead of an 802.11n wireless router.

Installation Costs

Installation costs are the fees or resources that are required to setup a product or service at a desired location. While many home media systems have auto-detection and automatic configuration capabilities, installation complications, such as interference with other devices or systems, might require a technician to install and configure a device, requiring a truck roll.

License Costs

A license is a contract that grants specific rights to the use of intellectual property. A license typically identifies issuers, principles, rights, resources, conditions and grants.

Home media networks use a mixture of technologies that may be protected by patent rights. The license fees that may need to be paid by manufacturers of HMN devices are likely to increase the costs of HMN equipment and systems.

Support Costs

Support costs are the combined expenses that are incurred to help customers and staff to setup and maintain products or services. When users add equipment or change their usage of devices on their HMNs, it can cause undesirable results. For example, if someone in a household begins to download movies (bit torrents), it can overwhelm the capabilities of the HMN, resulting in poor video quality. Consumers are likely to call the TV service provider to find a solution. The TV service provider will need to have technical support staff that can help to identify the challenges, which may not be the TV broadcasters fault, and recommend solutions.

Chapter 3

Home Networking Technologies

Home networking systems use key technologies to allow them to reliably distribute broadband transmission. These technologies include adaptive modulation, synchronized transmission, interference avoidance, power level control and channel bonding.

Adaptive Modulation

Adaptive modulation is the process of dynamically adjusting the modulation type of a communication channel based on specific criteria, such as interference or data transmission rate. Adaptive modulation allows a transmission system to change or adapt as the transmission characteristics change.

For some HMN systems, the characteristics of the transmission line can dramatically change over short periods of time. HMN systems typically share transmission resources (such as power lines) and the characteristics of these transmission lines can dramatically change. For example, light switches can form new wire path connections. Adaptive modulation allows the HMN transmission system to adapt to the new conditions that occur, such as slowing the transmission rate when interference occurs and returning to a more efficient transmission system when the characteristics of the transmission line change again.

Modulation Type

Modulation is the process of changing the amplitude, frequency or phase of a radio frequency carrier signal (a carrier) based on changes within the information signal (such as voice or data). In general, the more efficient the modulation type (more bits per change), the more sensitive the transmission becomes to distortion and interference. Modulation types that have limited level changes (for example tone on, tone off) are more robust and reliable to multiple level modulation systems (no tone, low tone, medium tone, high tone).

Figure 3.1 shows how adaptive modulation may be used to ensure reliable data transmission in changing conditions. This example shows a system that can adapt its modulation when it detects interference. This example shows two types of modulation; one that has two levels and another that has four levels. The modulation type that only has two levels is more robust that the modulation type that has four levels, which is more sensitive to interference and distortion.

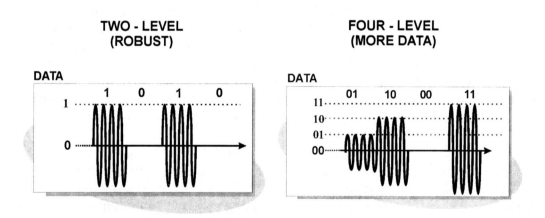

Figure 3.1, Adaptive Modulation

Echo Control

Echo control includes the processes used to identify echoes within transmitted signals, and the processes used to remove the distortions caused by signal echoes. For wired HMN systems, there can be many connection paths, some of which may not be properly installed or terminated, resulting in signal reflections.

Changes in the HMN systems may also occur as a result of the installation, control or removal of devices or changes in configurations. Endpoints that are not properly terminated, such as those that utilize handmade splices, can cause signal reflections, which can cause signal distortion.

Signal Reflections

Signal reflection is the sharp changing of the direction of a transmitted signal as it passes from one transmission medium to another (transmission channel or device). When the characteristics of the mediums are different (impedance), a reflected signal will be generated. Some of the energy from the forward signal (incident signal) is redirected (reflected) back towards the signal source. Echoed signals may be received at other devices, causing distortions in signal.

Echo Cancelling

The distortion caused by echoed signals in HMN systems can usually be removed through the use of echo cancellation. Echo cancellation is a process of extracting an originally transmitted signal from the received signal that contains one or more delayed signals (copies of the original signal). Echo canceling may be removed by performing advanced signal analysis and filtering. This echo cancelling process requires a home media network device (such as a powerline data adapter) to estimate the types of distortion and subtract the predicted distortion from the received signal.

Synchronized Transmission

Synchronized transmission is the transferring of information during a time period that is previously defined (time synchronized) or a time period that occurs after another event (such as a synchronization message). Synchronized transmission is used to ensure (guarantee) that certain types of information will be provided at specific, required times. Synchronized transmission may be used to provide certain types of services (such as video) and they may be coordinated by a master controller device.

Type of Service (ToS)

The use of synchronized transmission reserves transmission resources (bandwidth reservation) for time critical information such as television or telephone media. Bandwidth reservation is a process that is used to reserve bandwidth capacity through devices or communication lines for specific communication sessions or services. The use of synchronized transmission allows for reduced transmission delays (limited latency), minimizes jitter and provides low error rates.

Master Controller

An HMN system may designate one of its connected devices as a master controller. The master controller coordinates transmissions between devices that are part of the HMN. These devices request transmission resources (reserved bandwidth) from the master unit. If the master unit has the bandwidth available, it reserves the bandwidth by periodically sending messages throughout the HMN.

When an HMN is initially started (several devices are turned on), a master device is selected. Master device selection is the process used to select which communication node, such as a gateway device, will coordinate the other devices (slaves) in a network or system. The master slave relationship can be permanently or dynamically assigned. The dynamic assignment of a master slave relationship is necessary in communication systems where all devices provide similar functions.

Contention Free Transmission

To perform synchronized transmission, the master unit periodically sends media access plan (MAP) messages that contain a list of devices, as well as assigned transmission schedules and priorities for media transmission. After an HMN device has decoded the MAP message and has determined that it is time to transmit, it can begin transmitting without coordinating with other devices (contention free).

Contention Based Systems

Contention based systems allow devices to compete for transmission times with other devices within the system. These devices listen for periods of inactivity and, when these periods occur, transmission can begin. If interference from other devices is not detected at the start of transmission, a full packet of data can be sent.

Contention based devices and transmission may co-exist with contention free transmission. The contention based device may be able to receive information that identifies reserved (contention free) transmission times that are assigned to other devices. The contention device will then only compete for transmission time during authorized, non-contention free time periods.

Interference Avoidance

Interference avoidance is a process that adapts an access channel sharing process so that the transmission does not occur on specific frequency bandwidths. By using interference avoidance, devices that operate within the same frequency band and the same physical area can detect the presence of each other and adjust their communication system to reduce the amount of overlap (interference) caused. This reduced level of interference increases the amount of successful transmission, therefore increasing the overall efficiency and overall data transmission rate.

Interference Detection

Interference detection is the process of monitoring signal levels and their quality characteristics to determine if the correct signals are received in the appropriate format and sequence. Home media network systems may measure the number of bits that are received in error over periods of time to determine which channels or sub-channels are experiencing interference.

Channel Selection

Channel selection is the process of identifying, tuning or adapting and receiving a communication signal. Home media networks may be able to identify, request and setup one or several communication channels (or sub-channels) to use.

Network Coordination

Network coordination is the process of indentifying, analyzing and managing the use of channels (such as frequencies) in areas that have multiple networks to achieve acceptable performance for services within interference level limits.

Power Level Control

Power level control is the process of adjusting the power level of a transmitter on a communication link to minimize the amount of energy that is transmitted while ensuring that enough signal level is transmitted to achieve desired reception quality. In general, as the proximity of the transmitter to the receiver increases, the power level needed for transmission decreases.

Signal Level Detection

Power control is typically accomplished by sensing the received signal strength level and relaying power control messages between the receiver and transmitter, which indicate or instruct the transmitter to increase or decrease its output power level.

Power level control is used by some HMN systems to reduce the amount of signal leakage to a minimum. Because some HMN systems use high frequencies that travel over lines that are not capable of containing all of the signal energy (such as power lines), some of the energy leaks out. This signal leakage may cause interference with other devices such as AM radios.

Transmitter Power Control

Using signal level information from the receiving device, the transmitting device can adjust its transmission power level so only the necessary transmission level is used to ensure that a receiving device can successfully demodulate and decode its received signal.

Channel Bonding

Channel bonding is the process of combining two or more communication channels to form a new communication channel that can use and manage the combined capacity of the bonded transmission channels. Channel bonding can be used on HMN systems to increase the overall data transmission rate, and to increase the reliability of transmission during interference conditions.

Channel Multiplexing

Channel multiplexing is the separation of a signal channel into multiple sub channels. The data that is transmitted in a single channel (such as a video streaming channel) may be divided into segments (packets) that are sent over different communication channels.

Channel Combining (Demultiplexing)

Channel combining is inverse channel multiplexing. The channel combining process involves identifying and re-sequencing packets that are received on multiple communication channels back into a single channel.

Figure 3.2 shows how multiple transmission channels can be bonded together to produce a single channel with higher data transmission rates. This diagram shows how two transmission channels can be combined using a bonding protocol. This figure shows that a bonded session is requested and negotiated on a single communication channel. Once the bonded session has been setup, the bonding protocol is used to monitor and manage the bonded connection.

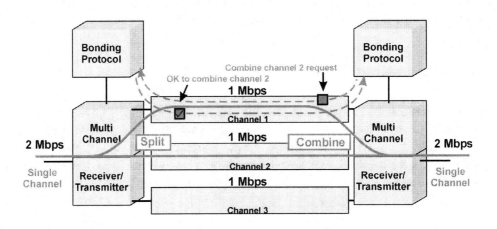

Figure 3.2, Channel Bonding

Topology Discovery

Topology discovery is a process where a network manager automatically searches through a range of network addresses, discovers devices and gathers information to discover the physical and logical relationships between nodes in a network which can be used to determine the layout and structure of a network. Topology discovery may be used to help a network optimize the communication paths and processes used in networks where more than one connection path can exist between devices, such as hybrid systems. Topology discovery involves link detection, node capabilities and network mapping.

Link Detection

Link detection is a process that monitors connection points for signal levels and commands to identify the presence of a communication line. Link detection may identify an active link (with a signal), along with its characteristics, which may include transmission types and protocols.

Node Capabilities

Node capabilities are the switching and processing functions that are available within a network connection point (such as a router).

Network Mapping

Network mapping is the identification of the interconnection between communication lines and nodes. It can be used to identify which paths may be used to connect devices within a home media network.

Chapter 4

Home Network Transmission Types

Home network transmission medium types for premises distribution include wired Ethernet (data cable), wireless, power line, phone line, coaxial cables and optical lines. Each of these transmission types has different capabilities and characteristics that may be suitable for some home media network requirements.

Wired LAN

Wired LAN systems use cables to connect routers and communication devices. These cables can be composed of twisted pairs of wires. Wired LAN data transmission rates vary from 10 Mbps to more than 1 Gbps.

When wired LAN systems use twisted wire pairs in the cable, the data transmission rate (cable rating) is based on the number of twists in the cable wire pairs. Cable wire pairs that have a higher number of twists can send data at higher transmission rates.

The frequency range of a wired cable is not the same as the data transmission rate that it can provide. More advanced modulation technologies can transfer several bits of information for each cycle (Hertz) of frequency that can be transmitted by the wire. This means that a cable that can transfer 250 MHz (such as category 6 Ethernet cable) can transfer up to 10 Gpbs of information.

A data cable that is used for wired LAN networks is classified by the amount of data it can carry in. This classification was established by The Electronics Industry Association/Telecommunications Industry Association EIA/TIA 586 cabling standard. In general, category 1 rated cable is unshielded twisted-pair (UTP) telephone cable (not suitable for high-speed data transmissions). Category 2 cable is UTP cable that can be used to provide transmission speeds up to 4 Mbps. Category 3 UTP cable can be used at data transmission speeds of 10 Mbps. Category 4 UTP cable can transmit up to 16Mbps and is commonly used for Token Ring networks. Category 5 cable is rated for data transmission speeds up to 100 Mbps. Category 5E (enhanced) has the same frequency range as Category 5, but provides a lower amount of signal transfer (crosstalk) so it can be used for 1 Gbps Ethernet systems. Category 6 cable has a frequency rating of 250 MHz and has demonstrated 10 Gbps transmission speeds.

Wired home LAN systems are typically installed as a star network. The star point (the center of the network) is usually a router or hub that is located near a broadband modem. LAN wiring is not commonly installed in many homes and when LAN wiring is installed, LAN connection outlets are unlikely to be located near television viewing points.

Wireless

A wireless local area network (WLAN) allows computers and workstations to communicate with each other using radio propagation as the transmission medium. The wireless LAN can be connected to an existing wired LAN as an extension, or it can form the basis of a new network. While adaptable to both indoor and outdoor environments, wireless LANs are especially suited to indoor locations such as office buildings, manufacturing floors, hospitals and universities. Wireless local area network systems (commonly called Wi-Fi) can distribute multimedia signals (such as IPTV).

Wi-Fi distribution is important because it is an easy and efficient way to get digital multimedia information where it is needed without the addition of new wires. Some consumers have refused to add new IPTV services due to rewiring or having to retrofit their homes to support it.

Figure 4.1 shows how an in-home Wi-Fi system can be used for home network (TV, telephone and data) distribution. This diagram shows that a broadband modem is installed in the home that has WLAN with premises distribution capability. This example shows that the broadband modem is located at a point that is relatively far from other devices in the home. The broadband modem is connected to a wireless access point (AP) that retransmits the broadband data to different devices through the home including a laptop computer, Wi-Fi television and a set top box (IP STB) that has a built-in Wi-Fi receiver.

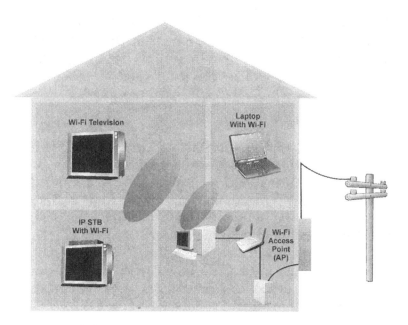

Figure 4.1, Wireless Home Media Distribution

Wireless transmission systems operate within defined frequency bands, have varying data transmission rates, and may include multibeam transmission capability.

Frequency Bands

Frequency bands are the range of frequencies that are used or allocated for radio services. The frequency bands that are commonly used for home media networks are in the unlicensed industrial, scientific and medical (ISM) band. The bands of the electromagnetic spectrum include frequency ranges of 902-928 MHz, 2.4-2.484 GHz, and 5.725-5.825 GHz, which do not require an operator's license. Most wireless LAN devices operate in the 2.4 GHz and 5.7 GHz band.

The requirements for unlicensed frequency bands (maximum power levels and transmission times) were designed to allow multiple devices to co-exist with each other with varying levels of interference. In general, there tend to be more devices that operate in the 2.4 GHz region than in the 5.7 GHz region, which means that interference may be higher in the 2.4 GHz band. It is possible to setup home networks to use 2.4 GHz wireless LANs for data transmission and 5.7 GHz wireless LANs for video distribution.

Data Transfer Rates

The data transfer rates that are available on wireless LANs vary based on modulation types, interference levels and channel bonding. Wireless LAN HMNs transfer user information over a WLAN system in a home or building. Wireless LAN data transmission rates vary from 2 Mbps to over 54 Mbps. Higher data transmission rates (up to 300 Mbps) are possible through the use of channel bonding (combining channels).

WLAN networks were not initially designed specifically for multimedia. In the mid 2000s, several new WLAN standards were created to enable and ensure different types of quality of service (QoS) over WLAN.

Multimedia signals such as television and music are converted into WLAN (Ethernet) packet data format and distributed through the home or business by wireless signals. Some versions of the 802.11 WLAN specifications include the ability to apply a quality of service (QoS) to the distributed signals, giving priority to time sensitive information (such as video and audio) to ensure that it can get through before non-time sensitive information (such as web browsing).

Figure 4.2 shows the different product groups of 802.11 systems and how the data transmission rates have increased in the 802.11 WLAN over time as new, more advanced modulation technologies are used. The first systems could only transmit at 1 Mbps. The current evolution of 802.11a, 802.11g, and 802.11n allows for data transmission rates of up to 54 Mbps.

	802.11 1	802.11 2	802.11A 3	802.11B 4	802.11G 5
Access Method	FHSS	DSSS	DSSS	DSSS	DSSS
Modulation Type	GFSK	DBPSK DQPSK	OFDM QPSK QAM	CCK	OFDM QPSK QAM
Frequency	2.4 Ghz	2.4 Ghz	5.7 Ghz	2.4 Ghz	2.4 Ghz
Data Transmission Rate	1-2Mbps	1-2Mbps	9-54Mbps	5.5-11Mbps	9-54Mbps

Figure 4.2, Wireless LAN Standards

Multi-beam Smart Antenna System

A multi-beam smart antenna system uses active transmission components to allow specific antenna patterns to be formed or selected. A smart antenna may have multibeam capability that allows for the reuse of the same frequency in the same radio coverage area. Using smart antenna systems, the transmission signal energy can be sent in a specific direction that does not result in interference with other signals that are operating in the same general area.

Wi-Fi systems that have smart antenna capability can alter multicast traffic by directing the traffic to a specific receiver, which forces that receiver to provide an acknowledgement. This way the system knows if the video transmission was received and the quality of the link.

Figure 4.3 shows how WLAN systems can be improved through the use of directional transmission and media prioritization to provide improved performance. This example shows that a wireless access point has been enhanced to allow the transmission of signals using directional antennas so that signals can be sent to specific devices. In this diagram, one of the best paths between the access point (AP) and Wi-Fi device (Wi-Fi Television) is not direct, as a metal art object that is located between the AP and the Wi-Fi device reflects the radio signal. This example also shows that this WLAN system has the capability to prioritize transmission based on the type of media (such as real time video over web browsing data).

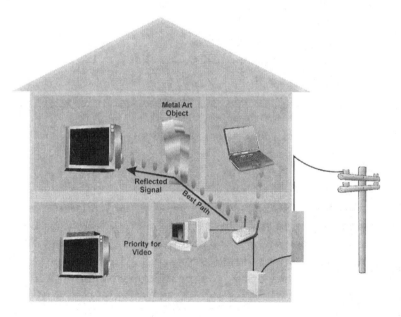

Figure 4.3, Smart Wireless LAN

Electrical Power Line Carrier (PLC)

An electrical power line carrier is a signal that can be transmitted over electrical power lines. A power line carrier signal is transmitted above the standard 50 Hz or 60 Hz power line frequency. Power line communication systems have evolved to overcome transmission challenges that include electrical noise, phase coupling and changing connection paths (topology).

The power line communication systems developed in the 1970s used relatively low frequencies such as 450 kHz to transfer data on power lines. The amount of data that could be transferred was limited and the applications typically consisted of controlling devices such as light switches and outlets. Some of the early home automation systems included X-10, CeBus and LONworks.

Electrical Noise

Electrical power lines are commonly exposed to undesired electrical noise signals that are created from a variety of sources such as brush motors, halogen lamps, dimmer switches and devices that produce noise signals that travel into the electrical lines. Electrical noise signals can be random and can result in the loss of data packets. Home media networks that use electrical power lines can use processes to detect and isolate many of the electrical noise signals.

Phase Coupling

Older (legacy) power line communication systems had challenges with wiring systems that used two or more phases of electrical power. The signals that traveled on wires for one phase of the electrical system would need to travel through other devices (such as 220 volt appliances) to reach wires that were connected to the other phase of electric.

Newer power line communication systems use higher frequency signals that can travel directly across (jump over) wires in an electrical panel so the entire home can distribute the home media network signals.

Changing Topology

The connection paths within an electrical distribution system can change over time. Connection paths may change when light switches are turned on (connected) or off (opened), and when appliances cycle on and off. These new connection paths may result in signal reflections that can cause distortion. Home media networks may be able to detect and rapidly adapt to the changing topology conditions of electrical wiring within homes.

Figure 4.4 shows how existing power line distribution systems in a home (such as X-10) can be used to distribute data signals. This diagram shows that medium voltage electricity (6,000 to 16,000 Volts) is supplied to a step down transformer near the home. This transformer produces two or three phases of power ranging from 110 V to 240 V. These electrical signals pass through an electrical distribution panel (circuit breaker panel) in the home

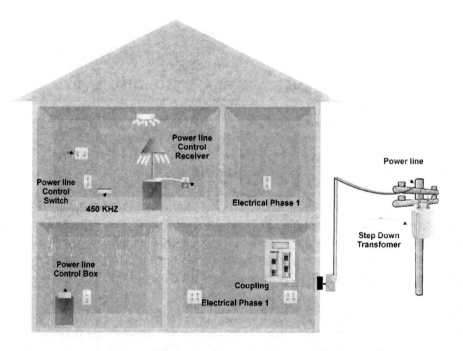

Figure 4.4, Power Line Communication Control System

to supply power to outlets and lights. This diagram shows an X-10 power line communication system where an X-10 switch in the wall controls an X-10 outlet receiver by sending control signals over the power line on the same circuit (same phase) to allow a user to control a lamp. This example also shows an X-10 power line control box plugged into another outlet on a different circuit (different phase) where the control signal must cross over from one phase to the other phase (cross over in the electrical panel) in order to reach the X-10 outlet receiver.

Although power line communication systems could technically transfer data in the 1970s and 1980s, improvements in the power line data transfer rates necessary for home data networking did not occur until the early 2000s. Part of the motivation to make advances in power line communication was the increased need for home networking.

Power line premises distribution for home media is important because televisions, set-top boxes, digital media adapters (DMAs) and other media devices are already connected to power outlets installed in homes or small businesses. These lines can be used to transmit rich multimedia content where it is desired.

Coaxial

Coaxial cable home media networks transfer user information over existing coaxial television lines in homes or buildings. Coaxial cable data transmission rates vary from 1 Mbps to over 1 Gbps. Many homes have existing cable television networks with outlets that tend to be located near video accessory and television viewing points. Coaxial home media distribution systems are commonly setup as tree structures, may co-exist with existing TV signals, and may use various types of coaxial cable.

Tree Distribution

Coaxial systems are commonly setup as tree distribution systems. The root of the tree is usually at the entrance point of the home or building. The tree may divide several times as it progresses from the root to each television outlet through the use of signal splitters.

Each time the signal passes through a splitter, the energy is divided between each of the connection paths (ports on the splitter). A splitter that has 4 ports will have at least a 75% reduction in signal power that goes to each port.

Cable Television Co-Existence

Home media networks that operate on coaxial lines may need to co-exist with the TV signals of existing television systems. This can be accomplish by transmitting the home media network signals on frequencies that are above the frequency range of the other television signals (typically above 850 MHz). This allows for the simultaneous distribution of existing TV signals, such as cable television channels, and home media network signals.

Coaxial Wire Types

There are different types of coaxial cable with characteristics that include frequency range and attenuation level. Older coaxial cable types (such as RG-59) tend to have lower frequency ranges and higher attenuation levels than newer types of cable (such as RG-62). In general, as the frequency of a signal increases, so does the amount of attenuation. Home media networks that use coaxial lines can adapt to this by sensing the signal attenuation levels at the receiving devices and increasing their transmitter power levels for higher frequencies.

Figure 4.5 shows how an in-home coaxial cable distribution system is typically used to distribute television signals in a home. This diagram shows that a cable connection is made at a demarcation (demarc) point on the outside of a home or building. This cable connects to a signal splitter and, optionally, an amplifier, which divides the signal and sends it to several locations throughout the home. This example shows that the coaxial systems may use lower frequency bands for return connections (data modems), middle frequency bands for analog television channels, and upper frequency bands for digital television channels.

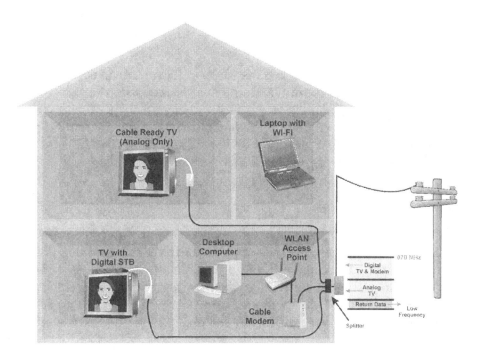

Figure 4.5, Coax Television Distribution system

Because the coax cable is shielded and RF channels are virtually free from the effects of interfering signals, coaxial cable provides a large information pipe that is capable of distributing multiple wide radio frequency channels. Coax is already installed in many homes, and television outlets are commonly located near media equipment such as televisions, VCRs and cable modems. Coaxial cable is easy to install and expand.

To overcome some of the loss that occurs within a cable distribution system, RF amplifiers are used, which can introduce distortion and block signals from home media network operation. If amplifiers are needed to strengthen the RF signal, they should be installed before the master control device so as to prevent disruption of the home media network.

Telephone Line

Home media networks may transfer user information over existing telephone lines in a home or building. Home media networks that use telephone lines can have data transmission rates that range from 1 Mbps to over 300 Mbps. Telephone line outlets may be located near television viewing points making it easy to connect adapters or viewing devices that have telephone line home media network connections.

Wire Types

Telephone lines to and from the telephone company may contain analog voice signals, uplink data signals and downlink data signals. These frequency bands typically range in size up to 1 MHz (some DSL systems go up to 12 MHz). Home media networks that use telephone lines may use frequencies above the 1 MHz frequency band to transfer signals to telephone jacks throughout the house so they do not interfere with existing telephone signals.

Adapter boxes or integrated communication circuits convert the video and/or data signals to high frequency channels that are distributed to different devices located throughout the house. To ensure the phone line signals do not transfer out of the home to other homes nearby, a blocking filter may be installed on the line that enters into the home.

The earliest telephone line home multimedia communication systems were used to allow computers to transfer files amongst each other (home data network), and to connect data communication accessories such as printers. These early telephone line data communication systems sent a limited amount of data using frequencies slightly above the audio frequency band.

These early systems had relatively low data transmission rates and they were fairly limited when compared to automatic telephone line communication systems.

As home media network signals travel down telephone lines, a portion of the signal is lost through the wires (absorbed or radiated). Signal frequency, the type of wire used, how the wire was installed and the length of the wire are all key factors in determining the amount of energy that is lost. Generally, as the length of a telephone line increases and the number of outlets increases, so does the amount of attenuation.

Interfering Signals

A challenge for transmitting data over the phone line is the presence of interfering signals and the variability in the characteristics of the telephone lines. Interference signals include telephone signals (ringing, DTMF, modems) and signals from outside sources, such as AM radio stations. Home media networks that operate on telephone lines may be designed to detect and avoid these signals by selecting sub-channels that do not have significant interference levels.

Splices

Phone line data communication systems need to work through existing telephone lines, which may be unshielded, that are interconnected in a variety of ways (looped or spliced throughout a house), and may be attached to a variety of telephone devices and accessories. Variability can be caused by poor installation of telephone wiring, telephone cords and changes in the characteristics of the telephone devices and accessories. To overcome the effects of interference and the variability of transmission lines, home media network systems can utilize adaptive transmission processes.

Signal Leakage

Because some of the energy leaks out of the telephone line, the maximum signal level authorized by regulatory authorities, such as the Federal Communication Commission, is relatively low. Home media networks may reduce transmitter power levels to the minimum levels necessary for devices within the system to receive a signal (power level control), reducing the amount of signal leakage.

Figure 4.6 shows how existing phone line distribution systems in a home or business can be used to distribute both voice and data signals in a home or building. This diagram shows that an incoming telephone line is directly connected to all telephone outlets in the home or building. In some cases, the lines are spliced together and, in other cases, the lines are simply connected in sequence. This example shows that a phone line data network uses frequencies above telephone line audio to distribute data signals by sharing the phone line.

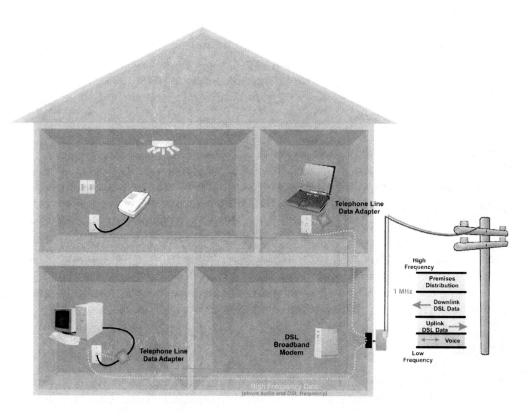

Figure 4.6, Phone line Communication System

Home Optical Networks

Some home networks, or portions of them, may transfer data over fibers in optical cables. The transmission rate in some optical cables can vary dramatically, based on the type of fiber and the type of optical signal source such as laser (very high) or LED (moderately high). The choices of fiber and optical transmitter can also determine the types of connectors and installation that may be required for optical systems. Typical data transmission rates for home optical networks range from 25 Mbps to 250 Mbps.

Fiber Type

Optical fibers can be designed and produced to transfer light in different ways. Single mode fibers transfer a single wavelength (color) of light through a very precise inner core, which is smaller than the size of a human hair. Multiple mode fibers can transfer multiple wavelengths of light through a much wider inner core. Because multimode fibers allow the light to bounce around within the core, the light pulses tend to become less precise or, more fuzzy, which limits the maximum data transmission rate and distance that can be used. However, the installation of multimode fibers with very large optical cores allows for the bending of fiber (around corners), which cannot be done with single mode fibers.

Optical Type

The transmitter for optical systems can use light amplification for the stimulation of radiation (LASER) or light emitting diodes (LED). Lasers create highly focused signals which can transfer data at very high speeds (10 Gbps+). LEDs tend create light signals that are more disbursed (less focused). The light energy from lasers may be enough to cause harm to people, especially to the eye. Home media networks that use optical lines may use LEDs instead of lasers to ensure that the system is safe to install and use.

Connection Type

Traditional optical connectors require precise connection points to avoid optical signal reflections and distortion. Optical systems that use low cost plastic fiber with relatively wide cores can have less precise connectors. The fiber cable can be cut by wire cutters or scissors and pushed into a spring loaded connector.

Chapter 5

Home Network Systems

Home media network systems (HMNs) are the combination of equipment, protocols and transmission lines that are used to distribute communication services in a home or building. HMNs may be used on one or several types of transmission lines such as coax, twisted pair, power line or wireless. Some of the more popular types of HMNs include HomePlug™, Digital Home Standard (DHS), HomePNA™, MoCA, 802.11 WLAN, and FireComms home optical systems.

HomePlug™

HomePlug is a system specification that defines the signals and operation of data and entertainment services that can be provided through electric power lines installed in homes and businesses. Development of the HomePlug specification is overseen by the HomePlug® Power line Alliance.

HomePlug 1.0

HomePlug 1.0 is a power line data communication system that transfers data through existing power lines at 14 Mbps using higher frequencies (from 4 MHz to 21 MHz). The HomePlug system uses a higher frequency range than previous power line data communication systems, which enables signals to couple phases in the electrical distribution panel. As a result, the coverage of HomePlug power line communication systems approached 99% without the need for cross phase couplers or professional installers.

As the signals travel down the power lines, a portion of the signal is lost through the wires (absorbed or radiated). Signal frequency, the type of wire, how the wire is installed and the length of the wire are key factors in determining the amount of energy that is lost. Generally, as the length of a power line increases and the number of outlets increases, so does the amount of attenuation. Signal loss at high frequencies through power lines can reach 50 dB, and the dynamic range of HomePlug devices can be between 70 dB and 80 dB.

Because it is possible for HomePlug signals to travel on power lines shared between several homes, the signals are encrypted to keep the information private. The HomePlug 1.0 system encrypts (scrambles) information using 56-bit DES security coding. Only devices that share the same HomePlug encryption codes can share the information transferred between the devices.

One of the significant challenges of power line communication systems is the sources and effects of interference signals that can distort power line communication signals. Interference signals include motor noise, signal reflections, radio interference, changes in electrical circuit characteristics, variability in the amount of coupling that occurs across different phases of electrical circuits, and stray signal transmissions. The HomePlug system was designed to overcome these types of interference and, in some cases, the HomePlug system can take advantage of them.

Motor noise is the unwanted emission of electrical signals produced by the rapidly changing characteristics of a motor assembly. In most homes, motors are in a variety of appliances and they may be used at different locations at any time. The HomePlug system can adapt in real time to the distortions caused by motors and appliances.

Signal reflection is the changing of direction traveled by a signal as it passes from one transmission medium to another (transmission channel or device). When the characteristics of the media are different (impedance), a reflected signal is generated. Some of the energy of the forward signal (incident signal) is redirected (reflected) back towards the signal source. When high-frequency HomePlug signals reach the ends of power lines, some of their signal energy is reflected back to the transmitter. These reflected signals combine with the forward (incident) signals, resulting in distortion. The HomePlug system performs a sophisticated analysis of the signal, and it is possible to use the signal reflections as an advantage rather than a challenge.

Some of the frequencies used by HomePlug systems are the same radio frequencies used by citizen band (CB) radios and AM broadcast radio stations. The HomePlug system divides the frequency band into many independently controlled sub-channels using a modulation scheme known as Orthogonal Frequency Division Multiplexing (OFDM) so that, when interference is detected (such as from a hair dryer), the sub-channels that are affected can be shut off. Because there are so many available sub-channels, this has little effect on the overall capacity of the HomePlug system.

Another common challenge with home power line communication is the dynamic change that can occur in electrical circuit characteristics as light switches are used and electronic devices are plugged into or remove from electrical outlets. The HomePlug system is smart enough to sense and adjust for the signal channels that occur due to changes in electrical circuit characteristics.

Each device that is part of an in-home HomePlug network requires a HomePlug adapter or a HomePlug converter to be built into the device. The HomePlug adapters convert information signals (digital data) into frequency carriers that travel down power lines. The adapters also coordinate access to the power line communication system by first listening to ensure that there is no existing activity prior to transmitting, and stopping transmission when they detect the occurrence of information packet collisions.

Figure 5.1 shows how the HomePlug 1.0 system allows a power line distribution system to transfer data between devices connected to electrical outlets in a home. This example shows several computers operating in a home and communicating between other data communication devices which are plugged into outlets. This example shows that a computer can send a data signal to a printer located at an outlet in another room. This diagram shows that the high frequency signals from the computer travel down the electri-

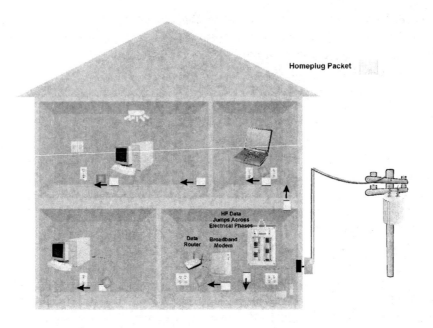

Figure 5.1, HomePlug 1.0 Distribution System

cal line to the electrical panel. Then, at the electrical panel to the power line communication signal, it jumps across the circuit breakers and travels down the electrical lines to reach the HomePlug 1.0 adapter that is connected to the printer. This figure also shows that the computer can send data signals down the power line to a HomePlug router that is connected to a broadband modem, allowing the computer to connect to the Internet.

HomePlug Audio Visual (HomePlug AV)

The HomePlug Audio Visual (HomePlug AV) specification was ratified in 2005 by the HomePlug Board of Directors, to provide home media networking. HomePlug AV was designed to give priority to media that requires time sensitive delivery, such as IPTV, while allowing reliable data communication, such as web browsing, to simultaneously occur. HomePlug AV uses a mix of random access (unscheduled) and reserved access (scheduled) data transfer. The Carrier Sense Multiple Access with Collision Avoidance (CSMA/CA) protocol provides for the efficient transfer of bursty data while the scheduled TDMA system ensures real time media (such as digital video and audio) will be delivered without delays, and will take priority on the wire over CSMA/CA traffic.

The HomePlug AV system is designed to be a 200 Mbps-class PHY device. The actual data transmission rate will be less than 200 Mbps due to channel access (MAC) efficiency needs, which are to be contemplated in overall performance measurements.

HomePlug AV uses a significantly higher spectrum than previous powerline systems (from 2 MHz to 28 MHz), which allows for good cross phase coupling as these higher frequencies can jump across circuit breakers to increase the signal availability in the home. Like HomePlug 1.0, the coverage of HomePlug AV power line communication systems approaches 99% without the need for cross phase couplers or professional installers.

The HomePlug AV system uses a more secure 128-bit AES encryption process to keep information private. The HomePlug AV system has better control of transmission delays (latency and jitter) and is designed to co-exist with the HomePlug 1.0 system. Because it operates at high frequencies, it does not interfere with power line control systems such as X-10.

The HomePlug AV system uses nearly 1,000 separate narrowband carriers. The system can selectively shut off some of these narrowband carriers (frequency notches) when it senses interference.

Like the HomePlug 1.0 system, each platform that will be part of the HomePlug AV network requires a HomePlug AV adapter or a HomePlug AV transceiver built into the device. The adapters coordinate access to the power line communication system. This coordination involves either reserving time periods for transmission and/or dynamically coordinating access by listening to the signals on the power line to ensure that there is no existing activity prior to transmission, and stopping transmission when data transmission collisions are detected.

Figure 5.2 shows how a HomePlug AV system can allow a power line distribution system to dynamically interconnect devices by connecting devices to electrical outlets and using high frequency signals to transfer information. This example shows that a HomePlug adapter coordinates the transmission of signals without interfering with the transmission between other devices. In this diagram, a TV service provider is sending a movie through a broadband modem to a television in the home. This example shows that the HomePlug system uses a HomePlug AV router to receive the signals from the broadband modem and route the information through the power lines to the television in the home using dedicated time slots to ensure a reliable quality of service (QoS). This figure shows that a computer is simultaneously connected through a HomePlug AV adapter to the in-home broadband modem, allowing it to connect to the Internet and transfer data on an "as available" basis.

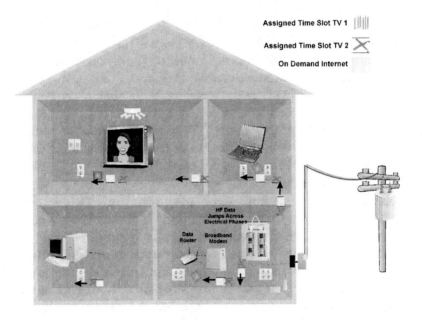

Figure 5.2, HomePlug AV Distribution System

It is possible for customers to self-install a HomePlug AV system. The system is setup to automatically detect (discover) other devices in the network. While most HomePlug devices are pre-configured with the most common settings, it is possible to customize the configuration of the HomePlug adapter devices by connecting them to a local computer and allowing the user to change key settings such as IP addresses and passwords.

Each device connected to a HomePlug system must have either an Ethernet or USB port, which allows a connection to an adapter, or a HomePlug AV transceiver embedded within a platform. The adapter can be an external box or it can be designed into a product, such as a CD player. Adapters can be simple conversion devices (Ethernet to HomePlug) or they can include some packet transfer capability (routers or bridges). A HomePlug Ethernet bridge allows packets from a HomePlug network to enter into an Ethernet network, such as allowing data to transfer into the Internet.

HomePlug devices include nodes and bridges. Nodes simply convert data signals into the frequency signals that travel on power lines. HomePlug bridges allow data packets to cross over into devices or other networks without any need to use configuration utilities.

The data transmission throughput is affected by the amount of overhead (control) data and the amount of information that is lost via interference and the access control process. Even with the overhead control information, the HomePlug AV system can have data transmission rates in excess of 100 Mbps. HomePlug AV technology is designed to co-exist with a variety of other systems including HomePlug 1.0, HomePlug 1.0 with Turbo, X-10, CeBus and LONworks.

The typical range of a HomePlug signal is approximately 1000 feet (300 meters). HomePlug technology generally works in most homes. A broad-based product ecosystem is rapidly forming for HomePlug AV technology. This ecosystem includes set-top boxes, routers, gateways, switches, displays, TVs, DMAs and entertainment electronics platforms. HomePlug devices that are certified to operate under the HomePlug specification will interoperate with each other.

Digital Home Standard (DHS)

Digital home standard (DHS) is a system specification that defines the signals and operation for data and entertainment services that can be provided through electric power lines that are installed in homes and businesses. The DHS system is designed with the ability to distribute signals over the medium voltage power distribution system (power lines within neighborhoods) and within the building (the home network).

DHS was designed to work with many different power and data system topologies worldwide. It can provide data transmission rates up to 200 Mbps [1]. Because the access system and home network share the same physical transmission lines, the access system and home network must share the 200 Mbps bandwidth.

The system is designed to allow for the co-existence of up to 3 in-home systems on the same local power system (within the same home). The DHS system has the capability to avoid interference with neighboring systems, such as those within neighboring homes.

The DHS system can be used for time sensitive streaming media. Its channel access latency for one system is less than 13 msec and the latency for three systems that co-exist is less than 38 msec [2]. To extend coverage, the DHS system can use frequency division repeaters or time division repeaters to receive and retransmit signals.

The DHS system is compatible with a wide area power line carrier (PLC) such as open PLC European Research Alliance (OPERA). Development of the DHS specification is overseen by the Universal Power Line Association (UPA). More information about DHS can be found at www.upaplc.org.

Figure 5.3 shows how the digital home standard allows for distribution over the power grid to the home and also distribution within the home. This diagram shows that the DHS system is designed to share capacity between the access portion of the system and within the home. The 200 Mbps data transmission rate is shared between the broadband power line (BPL) access network and the home network.

Chapter 5

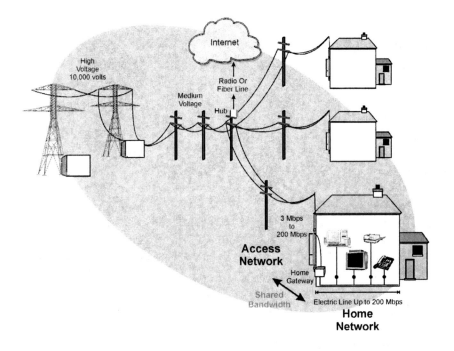

Figure 5.3, Digital Home Standard

HD-PLC

High definition power line communication is a communication system developed by Panasonic that uses high frequency signals over a power line to transmit data and digital media signals. The HD-PLC system transmits high-speed data signals using frequencies between 4 MHz and 28 MHz. The HD-PLC system supports the use of 128 bit encryption to ensure data privacy. HD-PLC has a maximum transmission distance of approximately 150 meters.

HomePNA™

The HomePNA is a non-profit association that works to help develop and promote unified information about home multimedia network technologies, products and services. HomePNA initially began with the definition of home data networks over telephone lines, and it has expanded its focus to include multimedia distribution over telephone and coaxial lines.

Sending broadband information over the same wires that are used to transfer telephone lines and coaxial lines involves converting the information into signals that can be transferred down these existing wires without interfering with other existing signals.

Phone line data communication systems use adapters to convert information from digital formats into signals that can travel along the telephone lines. Phone line data communication systems need to work through existing telephone lines (which may be unshielded) that are connected with each other in a variety of ways (looped or spliced throughout a house), and may have attachments to a variety of telephone devices and accessories.

The HomePNA system has evolved into a high-speed home multimedia communication network that can transfer digital audio, video and data over telephone lines or coaxial lines. The HomePNA system has been designed to co-exist with both telephone signals (lower frequency range) and digital subscriber line signals. Digital subscriber line (DSL) systems typically occupy frequency bands from the audio range up to approximately 2 MHz.

To connect to the HomePNA network, each device must have an adapter or bridge to convert the information or data into a HomePNA signal. Adapters or bridges can be separate devices or they can be integrated into a device such as TV set top box. The adapters convert standard connection types such as wired Ethernet (802.3), universal serial bus (USB), Firewire (IEEE1394) and others.

A gateway or bridge may be used to link the HomePNA network to other systems or networks (such as the Internet). The HomePNA system was designed to allow a gateway (such as a residential gateway - RG) to be located anywhere in the home.

A challenge for transmitting data over the phone lines is the presence of interfering signals and the variability of the characteristics of the telephone lines. Interference signals include telephone signals (ringing, DTMF, modems) and signals from outside sources, such as AM radio stations. Variability can be caused by poor installation of telephone wiring, telephone cords and changes in the characteristics of the telephone devices and accessories. To overcome the effects of interference and the variability of transmission lines, the HomePNA system uses an adaptive transmission system.

The HomePNA system monitors its performance, and it can change the transmission characteristics on a packet by packet basis as the performance changes. Variations can include changes in the modulation types and/or data transmission rates to ensure that communication between devices connected to the HomePNA system is reliable.

HomePNA 1.0

HomePNA 1.0 was the first industry specification that allowed for the use of data transmission over telephone lines. The HomePNA 1.0 system has a data transmission rate of 1 Mbps with a net throughput of approximately 650 kbps. HomePNA 1.0 systems are very low cost and are used throughout the world.

HomePNA 2.0

HomePNA 2.0 is a home networking system that is designed to provide medium speed data transmission rates and prioritize the transmission of different types of media. The HomePNA 2.0 system has a data transmission rate of 16 Mbps with a net data transmission throughput of approximately 8 Mbps. The HomePNA 2.0 system added new capabilities to home networking from the previous HomePNA 1.0 system, including prioritized transmission for different types of media, as well as an optional form of modulation to help ensure more reliable transmission during interference conditions.

Transmission prioritization was added to the HomePNA system to allow for differentiation and prioritization of packets for time sensitive services. These prioritized services included IP telephone and video services where longer packet transmission delays were not acceptable.

The HomePNA 2.0 system included the ability to transmit data using a modulation scheme known as Frequency Diversity Quadrature Amplitude Modulation (FDQAM). When interference is detected, such as from an AM radio station, the modulation type can change from QAM to FDQAM, increasing the robustness (reliability) of the signal. The FDQAM performs spectral multiplication where the same signal is transmitted on different frequencies (frequency diversity).

HomePNA 3.0

HomePNA 3.0 is a multimedia home networking system that is designed to provide high-speed data transmission rates and to control the quality of service for different types of media. The HomePNA 3.0 system has a data transmission rate of 240 Mbps with a net data transmission throughput of approximately 200 Mbps, and is backward compatible with HomePNA 2.0.

HomePNA 3.0 was standardized in June 2003 and commercial products were available starting in May 2004. The industry standard has been adopted by the ITU as specification number G.9954.

HomePNA 3.0 uses either quadrature amplitude modulation (QAM) for high-speed transmission or frequency diverse quadrature amplitude modulation (FDQAM) based on the quality of the transmission line and the media transmission requirements. The data transmission rate is constantly adjusted on a packet by packet basis by varying the symbol rates (increasing and decreasing the amount of shifts per second) and the number (precision) of decision points per symbol to ensure that data can be received in varying conditions.

The HomePNA 3.0 system added the ability to mix controlled (contention free) and on demand (random access) transmission within home networking. The HomePNA 3.0 system accomplishes this by assigning one of the HomePNA devices in the system as a master control unit. The master unit coordinates the transmission time periods of all of the HomePNA devices in the system through the periodic transmission of a media access plan message. A media access plan is a list of devices and their assigned transmission schedules and priorities for media transmission. The master unit coordinates all of the bandwidth reservation and transmission assignments, and all the other units are slaves that follow the master's lead.

The master can be selected automatically, or it can be manually setup or programmed as the master. For example, a TV service provider can setup the residential gateway (broadband modem) to be the master unit of the HomePNA system.

Each HomePNA 3.0 device is typically capable of acting as a master or a slave. If the master unit is shut off or if it becomes disabled, the HomePNA system will automatically setup another unit as the master.

Figure 5.4 shows a HomePNA 3.0 system. This diagram shows that the master unit periodically sends MAP messages. Each MAP message identifies which devices are assigned synchronized transmission, and when unscheduled transmissions may occur. This example shows that scheduled devices only transmit during their assigned schedules and other devices compete for access during the remainder (unscheduled portion) of the media cycle.

HomePNA 3.1

HomePNA 3.1 is a multimedia home networking system that is designed to provide high-speed data transmission rates and to control the quality of service for different types of media over both telephone lines and coaxial cables. The HomePNA 3.1 system has a data transmission rate of 320 Mbps with a net data transmission throughput of approximately 250 Mbps. The HomePNA 3.1 system is backward compatible with HomePNA 3.0 and HomePNA 2.0

Home Media Networks

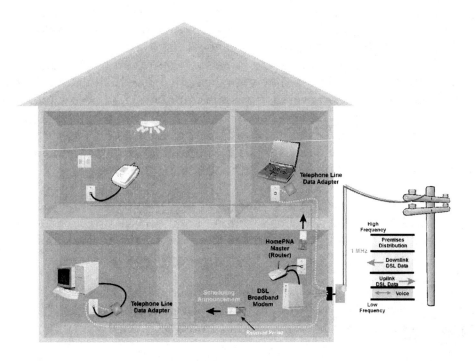

Figure 5.4, HomePNA 3.0 Distribution System

The HomePNA 3.1 system uses the same frequency band for telephone lines and coaxial lines. This frequency band is above digital subscriber line signals and below television channels. When operating on the coaxial line, the HomePNA 3.1 system only uses the scheduled mode which can be more than 90% efficient, increasing the data rates available for users. Because the coaxial line is a shielded line, the HomePNA system can increase its power, which is referred to as a power boost, allowing the HomePNA system to operate at an effective distance of over 5,000 feet (1,500 meters).

Figure 5.5 shows how a HomePNA 3.1 system can use both telephone lines and coaxial lines. The HomePNA system allows devices to directly communicate with each other. This diagram shows that the HomePNA system can be used to distribute voice, data and video signals over both telephone lines and coaxial lines, allowing it to be available at more locations in the home.

68

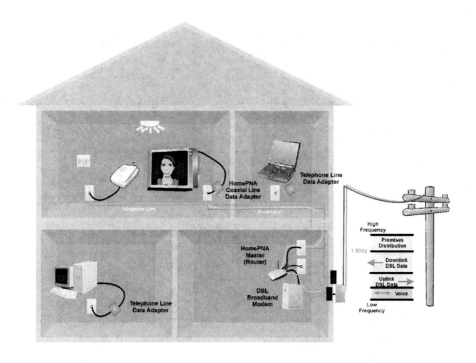

Figure 5.5, HomePNA 3.1 Phone line and Coax Distribution

The typical range of a HomePNA signal is approximately 1,000 feet (300 meters) when utilizing telephone lines. HomePNA devices that are certified to operate under the HomePNA specification will interoperate with each other.

Some telephone devices and accessories that are of lower quality can create interference with the HomePNA system. While this interference would typically only reduce the performance (data transmission rate), installing microfilters to block the unwanted interference from telephone devices can eliminate it.

Figure 5.6 shows a summary of the different types of HomePNA systems and their capabilities. This table shows that the HomePNA system has evolved from a low speed data only to a high-speed multimedia multiple line type home networking system.

System	Data Rate	Type of Media	Notes
HomePNA 1.0	1 Mbps, (650 kbps throughput)	Data Only	QAM Only
HomePNA 2.0	16 Mbps (8 Mbps throughput)	Prioritized for Multimedia	FDQAM and QAM
HomePNA 3.0	240 Mbps (200 Mbps throughput)	Scheduled Guaranteed QoS for Digital TV	FDQAM and QAM
HomePNA 3.1	320 Mbps (250+ Mbps throughput)	Scheduled Guaranteed QoS for Digital TV	Telephone line and Coax

Figure 5.6, HomePNA Capabilities

Multimedia over Coax Alliance (MoCA)™

Multimedia over Coax Alliance is a non-profit association that works to help develop and promote unified information about networking technologies, products and services that are primarily distributed over coax cabling systems within a building or facility.

The MoCA system operates by transferring media on frequency bands between 860 MHz and 1.5 GHz using the coaxial cable system [3]. The MoCA system uses 50 MHz transmission channels that transmit at 270 MHz. Each 50 MHz channel is subdivided into smaller channels using orthogonal frequency division multiplexing (OFDM), and the net throughput of each MoCA channel is 135 Mbps. Because the MoCA system can have multiple devices and it may co-exist with other systems, it can use up to 16 different channels that can be dynamically assigned and controlled. These channels can operate independently allowing MoCA systems to co-exist together.

The MoCA system is designed to prioritize the transmission of media to ensure a reliable quality of service (QoS) for various types of media. The MoCA system can schedule (reserve) bandwidth so it can provide different QoS levels. The MoCA system uses the data encryption standard (DES) for link level encryption to ensure privacy between neighboring devices and systems.

The MoCA system was designed to operate with most existing home coaxial networks without requiring changes. Home coaxial networks are commonly setup as a tree structures, and there may be several signal splitters used within the tree. To operate without any changes, signals from one device in the MoCA system must travel across splitters, a process which is referred to as splitter jumping. The insertion loss (attenuation) can be high when a different output port of the splitter is involved.

A large link budget allows for splitter jumping within the MoCA system. Link budget is the maximum amount of signal losses that may occur between a transmitter and receiver to achieve an adequate signal quality level. The link budget includes splitter losses, cable losses and signal fade margins.

A potential challenge for bi-directional coaxial distribution systems is the use of coaxial amplifiers. Coaxial amplifiers are devices or assemblies that are used to amplify signals that are transmitted on coaxial cables. Coaxial amplifiers may offer unidirectional or bidirectional amplification capabilities. The frequency range and coaxial amplifiers are typically chosen to match the signals that are transmitted on the coaxial cable.

Coaxial amplifiers may be used on home television systems with long cable connections or many television ports (many splitters). There are two types of coaxial amplifiers; drop amplifiers and inline amplifiers. Drop amplifiers are signal amplifiers that have been inserted after a communication drop within a building, but prior to the connection of other devices with the premises network. Inline amplifiers are signal amplifiers that have been inserted into a premises network at location in between (inline with) devices. Inline amplifiers can be a potential block for coaxial distribution signals. In the United States, about 2% of homes have inline amplifiers, and approximately ½ of these homes may have challenges using an HMN system [4].

Figure 5.7 shows how a MoCA system utilizes a coaxial distribution system to dynamically interconnect devices using coax and frequencies above 860 MHz. This example shows that the MoCA system is designed to operate in an existing cable television system by allowing signals to jump across and through existing splitters.

Home Media Networks

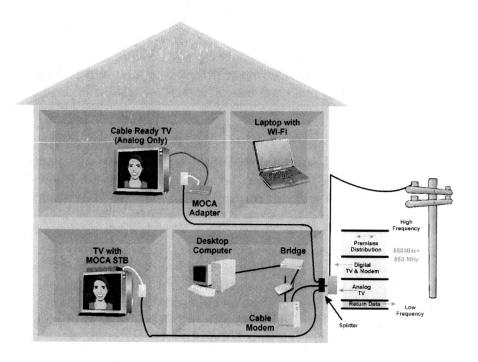

Figure 5.7, MOCA Distribution System

802.11 Wireless LAN

802.11 Wireless LAN, commonly referred to as Wi-Fi, is a set of IEEE standards that specify the protocols and parameters that are applied to commonly available wireless LANs that use radio frequency for transmission. The use of Wi-Fi systems for home media distribution offers the advantage of requiring no new wires. Wireless LAN systems operate in unlicensed frequency bands.

Chapter 5

Quality Performance Monitoring

Wi-Fi is fundamentally an unreliable medium due to that fact that it is shared among multiple users over the unlicensed RF spectrum, which allows all users to interfere with one another. The key challenges faced by providers and subscribers running real time media, such as telephony or IPTV, over Wi-Fi include radio signal interference from other sources, signal quality levels, range and data transmission rates.

The use of unlicensed frequency bands can also result in interference from other types of devices that use the same frequency bands. This includes interference that may result from other Wi-Fi systems operating nearby, microwave ovens or other obstructions that may dynamically appear and cause problems in the transmission of video over Wi-Fi.

Figure 5.8 shows typical types of unlicensed radio transmission systems that can cause interference with WLAN systems. This example shows that there are several different communication sessions simultaneously operating in the same frequency band, and that the transmission of these devices is not controlled by any single operator. These devices do cause some interference with each other and the types of interference can be continuous, short-term intermittent or even short bursts. For the video camera (such as a wireless video baby monitor), the transmission is continuous. For the cordless telephone, the transmission occurs over several minutes at a time. For the microwave oven, the radio signals (undesired) occur for very short bursts only when the microwave is operating. For the wireless headset, the transmission occurs for relatively long periods of time but the power is very low so interference only occurs when the wireless headset gets close to WLAN devices.

Radio interference can cause lost packets. This results in sub-standard video that freezes or has an unacceptable amount of artifacts. Sources of WLAN interference are varied and many including microwave ovens, baby monitors, cordless telephones, neighboring Wi-Fi networks and even people.

In addition, to overcome potential interference issues, Wi-Fi systems that are used for TV systems must be able to support quality of service for different types of media. If they cannot, a user watching a TV stream might

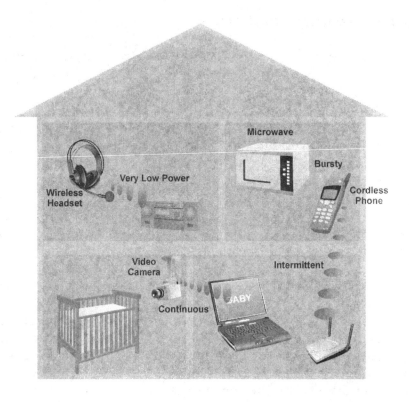

Figure 5.8, Wireless LAN Interference

have their viewing interrupted when another user is downloading a file over the same Wi-Fi network. Wi-Fi systems that are used for TV multimedia home networks may use a combination of 802.11e quality of service, 802.11n MIMO and smart antenna systems to increase the reliability and performance of the Wi-Fi system.

802.11e Quality of Service

The 802.11e wireless local area network is an enhancement to the 802.11 series of WLAN specifications that added quality of service (QoS) capabilities to WLAN systems. The 802.11e specification modifies the medium access control (MAC) layer to allow the tracking and assignment of different channel coding methods and flow control capabilities to support different types of applications such as voice, video and data communications.

Enhanced Distributed Channel Access (ECCA)

Enhanced distributed channel access is a medium access control (MAC) system that is used in the 802.11e WLAN system to enable the assignment of priority levels to different types of devices or their applications. The prioritization is enabled through the assignment of different amounts of channel access coordination back-off times, which are dependent upon packet priority. EDCA defines four priority levels (four access categories) for different types of packets.

To coordinate and prioritize packets between devices within an 802.11 network, a hybrid coordination function controlled channel access (HCCA) can be used. The HCCA uses a central arbiter (access coordinator) that assigns the access categories for different types of packets. The central arbiter receives transmission requests from all the devices communicating in the 802.11 system, so it can assign and coordinate transmission and tasks for other devices. The HCCA process can guarantee reserved bandwidth for packets classified by the EDCA process.

802.11n Multiple Input Multiple Output (MIMO)

The 802.11n wireless local area network is an enhancement to the 802.11 series of WLAN specifications that adds multiple input and multiple output (MIMO) capability to WLAN systems. The 802.11n specification modifies the medium access control (MAC) layer to allow for channel bonding (channel combining). This increases the available data transmission rate, increasing the reliability (robustness) of the Wi-Fi system.

Multiple input multiple output is the combining or use of two or more radio or telecom transport channels for a communication channel. The ability to use and combine alternate transport links provides for higher data transmission rates (inverse multiplexing) and increased reliability (interference control).

Figure 5.9 shows how a multiple input multiple output (MIMO) transmission system delivers signals over multiple paths to a receiver where they are combined to produce a higher quality signal. This example shows that, even if an object or signal distortion occurs in one of the transmitted paths, data can still be transmitted on alternative paths.

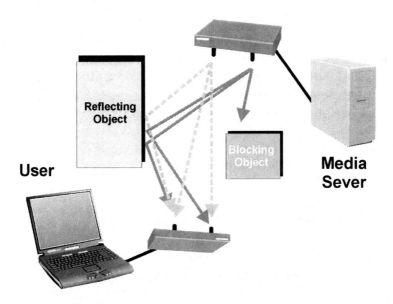

Figure 5.9, Multiple Input Multiple Output (MIMO)

HomeGrid

HomeGrid is a multimedia home networking system that is designed to provide high-speed data transmission rates and control the quality of service for different types of media that can be sent over different types of transmission lines (telephone lines, power lines and coaxial lines).

Unified Network

The HomeGrid system allows for the co-existence and potential combined use of multiple transmission types including copper telephone lines, electric power lines and coaxial lines. The HomeGrid system can use device profiles to help ensure that services, and the devices that use them, will operate in a specific way (resulting bandwidth, packet loss rate and packet delay) when providing services or performing applications. The unified network may coordinate transmission formats such as channel selections, frequency bands and transmission window sizes.

Home Media Domains

The HomeGrid system defines domains (groups of managed devices) within home network systems. Each domain is controlled and coordinated by a domain master (DM) and multiple domain masters are controlled and coordinated by a global master (GM). Relay nodes within a domain receive and forward packets to other parts of the network. Domain access points allow media and data to enter and leave the HomeGrid domains.

Foreign Networks

The HomeGrid system attempts to identify foreign networks and adapt its system to co-exist with such networks. The HomeGrid network can attempt to identify the carrier schemes that are used by other networks (alien networks) and communicate with them using Inter-system protocol. If the HomeGrid system is able to coordinate with other networks, both systems can be adapted. If the HomeGrid system cannot coordinate with other networks, it may attempt to adapt its own transmission scheme to minimize interference between the systems (such as inhibiting or masking the use of sub-carriers).

Home Media Networks

Figure 5.10 shows how the HomeGrid system can mix multiple types of transmission technologies using multiple domains. This example shows that each home media system is coordinated by a domain master (DM) and the entire system is coordinated by a global master (GM). The global master can also help to coordinate the operation of the HomeGrid system with other networks (Alien networks) by choosing and adapting the transmission bands and subchannels.

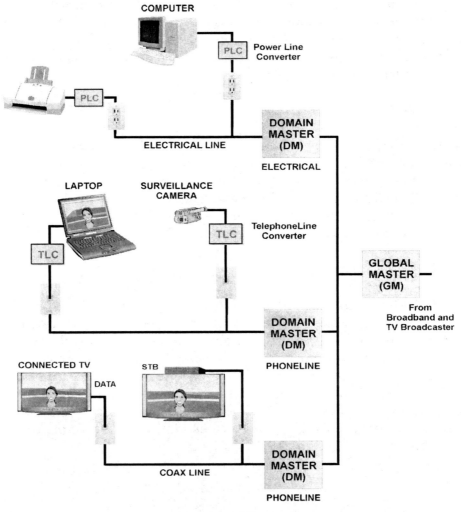

Figure 5.10, HomeGrid System

FiberComms Home Optical Network

FiberComms produces an optical home networking system that is designed to provide high-speed data transmission rates within the home over lightweight and easy to install plastic fiber cable, which uses low light energy light emitting diode (LED) transmitters. The Fibercomms optical lines have a maximum distance of over 100 meters.

POF is already in mass deployment in several industries including automotive and industrial networking. As of early 2009, POF has already been utilized in more than 15 million cars to interconnect audio, navigation and computer control systems via lightweight, EMI-immune, low cost lines.

LED Optical

The Firecomms optical home network uses low power light emitting diodes (LEDs) which can be safely viewed by the human eye. Since the data signal can be seen without the need for special equipment, maintenance costs are lowered, and troubleshooting is fast and easy.

Plastic Fiber

The Firecomms optical home network uses plastic optical fiber (POF) which is thin (2 to 4 mm thick), cost effective and easy to self-install. POF can be installed without the need for special tools or training. The POF cable is also thin and lightweight, so transport and handling costs are minimized.

All types of fiber optic lines work by focusing and directing light signals to travel down a clear optical channel (fiber light guide). When the light hits the side of the fiber, it is reflected or refracted (bent) back toward the center of the fiber. This process repeats until the light emerges out of the other end of the fiber.

Optical home networks use plastic fibers instead of the glass fibers that are typically used in wide area and long distance networks. Fibers commonly used in long distance telecommunication systems use single mode fiber (SMF), which lets only one narrow type of optical signal through. The width of the POF fiber core (where the light travels) is a millimeter wide as compared to the core of an SMF, which is 100 times smaller.

The optical transmission portion of the plastic fiber, referred to as the core, is relatively wide when compared to glass fiber. This allows the home optical fiber to be bent around objects, which is not possible with single mode glass fibers.

Plastic optical fiber lines contain almost all of the signal energy within the center core of the fiber. This means that fiber does not emit RF signals like copper cable, so it will not interfere with the existing infrastructure.

Figure 5.11 shows some of the differences between glass and plastic fiber transmission systems. The glass fiber transmission system uses a highly focused light source (laser) which transfers a light signal through a narrow strand of glass fiber. The plastic fiber transmission system uses a light emitting diode (LED) light source which transfers some of its light signal through a strand of plastic fiber.

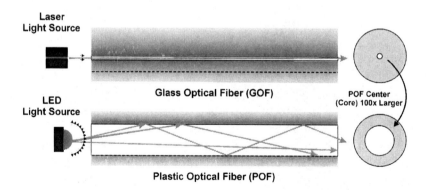

Figure 5.11, Glass and Plastic Fiber Transmission Differences

The thin POF cable can be easily installed under baseboards and along carpet edges or around door moldings. It can be pulled under carpets, fit through small holes, or run inside the wall cavity and attic, as is done with traditional wiring practices.

Opto-Lock Connectors

The Firecomms system uses OptoLock® spring loaded connectors. To connect the fiber, it is cut to the required length with standard cutters or scissors. Plastic optical cable has two strands; one for transmission, and the other for reception between devices. One can easily separate these fiber strands by pulling them apart. The fiber strand without the light is placed into the optical connector hole with the light. The fiber strand with the light is placed into the hole without the light. The sliding lock on the connector is pressed and the connection is complete.

References:
[1]. "OPERA, Broadband Over Power Line (BPL)", 2007-09-21, www.technologynewsdaily.com.
[2]. "Powerline Communication Systems - Access/In-home & In-home/In-home coexistence mechanism - General specifications", Version 1.0, Published by the Universal Powerline Association, 27 June 2005, www.upaplc.org.
[3]. "IPTV Crash Course," Joseph Weber, Ph.D. and Tom Newberry, Mc-Graw-Hill, New York, 2007, pg. 172.
[4]. "MOCA" Inline amplifiers.

Chapter 6

Home Media Management

Home media management is the processes that are used to identify, select, adapt and control media distribution within a home or facility.

There are many types of devices within the home. Some are media players, some are media recorders, and others can be used to control the selection or transfer of media between devices. Some manufacturers work together to define how the devices should operate and communicate with each other. Some of the home media management systems include DLNA and UPnP.

Digital Living Network Alliance (DLNA)

The Digital Living Network Alliance is a group of companies that work to create sets of interoperability guidelines for consumer devices in the home. The DLNA has defined a system of media formats, protocols and processes that are used to select, transfer and manage digital media devices that operate within the home. More information about DLNA can be found at www.DLNA.org.

A key goal of DLNA is to provide for interoperability between multiple types of devices and media formats. The DLNA system can use existing protocols including UPnP AV and UPnP printer (media management), HTTP and RTP (media transport), Internet protocol (network connections) and Ethernet, 802.11 WLAN, and Bluetooth (physical connections). To overcome incompatibilities between media formats, device types and data connection capabilities, the DLNA system may adjust its operation and features.

DLNA divides consumer devices into key types of domains that include computers, mobile and consumer electronics. DLNA defines devices that can process media as home network devices (HNDs). HNDs can be divided into classes, which include media servers, controllers, players, renderers and printers.

Digital Media Server (DMS)

A digital media server is a computing device that can process requests for and deliver digital media. A DMS may perform the acquisition, storage and transfer of media content such as videos, pictures and audio files. DMS functions may be included in advanced set top boxes, digital video recorders (DVRs) and digital tuners.

Digital Media Controller (DMC)

A digital media controller is a device or software application that discovers and coordinates access to media on digital media servers and directs it to digital media rendering (DMR) devices. Examples of DMCs include televisions, computers and interactive remote controls.

Digital Media Player (DMP)

A digital media player is a device or a software application that can request and receive media such as video, audio or images and convert it into a form that can be experienced by humans. DMP devices include personal computers with media player software, television monitors and multimedia mobile telephones.

Digital Media Renderer (DMR)

A digital media renderer is a device or software application that can display media to a viewer after the media has been processed or made available by another device. DMR devices include video monitors, digital displays and audio speakers.

Digital Media Printer (DMPr)

A digital media printer is a device or a software application that can convert media into a form that can be transferred to printed formats. Examples of DMPrs include photograph printers, inject printers and laser printers.

Figure 6.1 shows an example of the key components of a DLNA system within a home network. The digital media server (DMS) stores movies and audio files, which can be accessed by other devices that are connected to the home network. A digital media controller (DMC) is used to find and control the transfer of media within the home network. Several devices in the home network include a digital media player (DMP) that allows the user to play media. A digital media renderer (DMR) converts the media into a format that can be used to transform media into another format. A digital media printer (DMPr) converts media into a printed format.

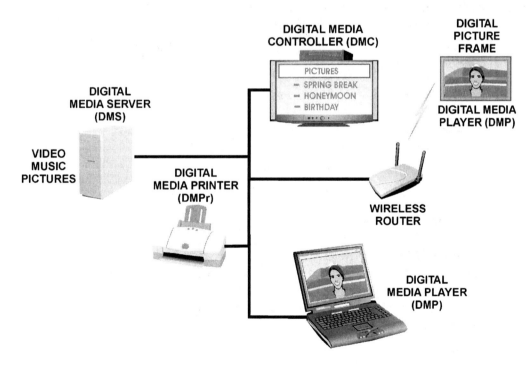

Figure 6.1, DLNA System Components

DLNA Layers

The DLNA system is divided into functional layers that include media format, media management, device discovery, media transport, network stack and physical connection layers.

Media Format Layer

The media format layer is a collection of processing functions that enable the conversion (transcoding) of media formats. An example of media conversion is the changing of AAC audio coding into MP3 audio coding, or the conversion of WMV9 video coding into MPEG-2 video coding.

Media Management Layer

The media management layer is a collection of processing functions that enable the identification, selection and control of media items. The DLNA media management layer implements the universal plug and play audio visual (UPnP AV) standard.

Device Discovery and Control Layer

Device discovery and control layer is a set of processes that are used to identify, select and allow interaction between devices. For the DLNA system, the device discovery and control layer uses the universal plug and play (UPnP) device architecture.

Media Transport Layer

Media transport layer is the commands and processes that are used to setup and coordinate the transfer of media over connections or through systems. The primary purpose of the transport layer is to provide reliable transfer of data between end users. For the DLNA system, the transport layer uses HTTP protocol.

Network Stack Layer

Network stack layer is a collection of processing functions that coordinate the communication between devices within a network. The network stack layer adds the address and control information to allow packets to travel through the network and find their destination.

Connectivity Layer

Connectivity layer is the set of processes and functions that adapt data or media into the physical connection channels from a device. DLNA connectivity options include 802.3 wired Ethernet, 802.11 wireless Ethernet, Bluetooth and other connection types.

Figure 6.2 shows how DLNA layered architecture can allow different types of devices to interact with each other. This example shows how a personal digital assistant can be used to control and select media that will be transferred to the television. The PDA communicates with the media format layer to convert its stored media into a format that can be displayed by the television. The media management layer allows for the media stored within the PDA to be identified and transferred to the television. The device discovery and control layer allows the PDA to find other devices (such as the television) to interact with. The media transport layer oversees the division of media into data packets that can be transferred between the PDA and the television. The network stack layer assigns addresses to each packet so they can travel from the PDA through the home network to reach the television. The connectivity layer allows for the conversion of data packets into wired or wireless Ethernet packets that travel through the network.

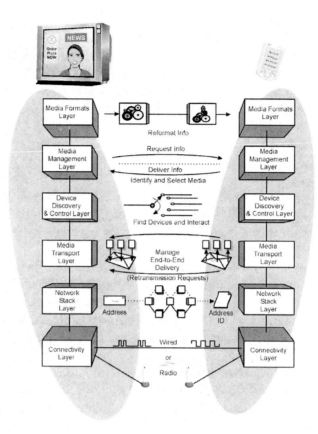

Figure 6.2, DLNA Layered Architecture

Universal Plug and Play (UPnP)

Universal plug and play is an industry standard that simplifies the installation, setup, operation and removal of consumer electronic devices. UPnP includes the automatic recognition of device types, communication capabilities, service capabilities, activation and deactivation of software drivers and system management functions. More information about UPnP can be found at www.UPnP.org.

Automatic Addressing

Automatic addressing is a process that dynamically assigns a label that can be used to identify and connect to a device that is attached to a system or network. UPnP devices use dynamic host configuration protocol (DHCP) to initially obtain an IP address when they are first connected to a device or system (AutoIP). If a DHCP server is not available, the UPnP device will assign itself a unique IP address, which allows it to communicate with the device it is connected to. If the domain name server has domain name capability and assigns it to the device, the device will use its domain name rather than its IP address when communicating with other devices or systems.

Device Discovery

Device discovery is the detection of a new device connection. UPnP enabled devices automatically sense when a connection has been changed (added or removed). When a connection has been detected, the device can advertise (broadcast) its available service information to other devices that it is connected to using simple service discovery protocol (SSDP).

Capabilities Description

UPnP devices include descriptions of their capabilities (functions and services). These capabilities can be exchanged with other devices in extensible markup language (XML). This information can include model identification, service capabilities and a list of other embedded devices and services.

Control Functions

The UPnP system defines some points as control points. Control points can send and receive command information from other devices that are attached to the system. Control messages are sent using the simple object access protocol (SOAP). When UPnP devices fist become attached, they attempt to find one or more control points.

Event Notifications

Event notifications are messages that are sent after a defined condition has occurred. UPnP devices may subscribe to other devices to receive event notifications. Event notification processes are defined by the general event notification architecture (GENA). Events may include information related to the condition, such as device state variables.

Presentation

Presentation is the portion of a system that converts media or data into a format that can be viewed or used by a person. UPnP devices may have a URL address assigned for the presentation of content, allowing a rendering device to stream or transfer content so it can be presented on a device.

Figure 6.3 shows an example of how UPnP can be used to attach a printer to a computer and how it can be accessed by a home network. This example shows that when the printer is first connected to a computer USB port, it

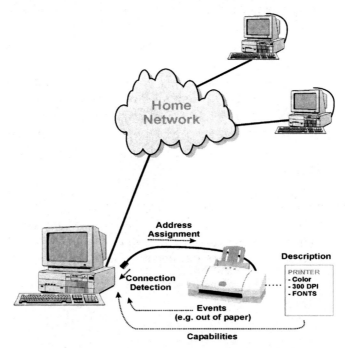

Figure 6.3, Universal Plug and Play (UPnP)

automatically obtains an IP address from the network (automatic addressing). The device then uses simple service discovery protocol (SSDP) to advertise (broadcast) its availability to the computer and other devices connected to the network (the printer is available). The host device (the computer) then requests service and capabilities information from the printer (color capability, paper size, fonts). Control commands may be exchanged between the computer and the printer (such as document print). Event notifications (unscheduled changes in conditions) from the printer may be sent to the computer when there is a status change, such as when the printer runs out of paper. Presentation processes allow for the adaptation and formatting of media to the display or player capability of the device such as the display of a "paper out" status on the computer monitor.

UPnP Audio Visual (UPnP AV)

Universal plug and play audio visual (UPnP AV) is an evolved version of the UPnP industry standard that adds new multimedia capabilities and device classes to the UPnP system. An UPnP AV system device includes a media server, media control points, media rendering, rendering control, quality of service and remote access.

Media Server

UPnP media servers are storage and processing devices that can store, transfer (stream), and provide media library information to other devices. UPnP media servers may be hardware or software based. Hardware based media servers are specifically designed for media storage and streaming. Software based UPnP servers may run on several types of computer servers, such as Linux, Windows, Mac OS X.

Control Point

Control points can automatically discover, connect and coordinate the operation of media servers and other devices. Control commands are expressed in simple text database formats (XML), which are transferred using simple object access protocol (SOAP).

Media Renderer

A media renderer is a slave device, one that runs under the control of a master control point, which can covert media into formats that can be viewed or used by users.

Rendering Control

Rendering control is the ability to send and respond to media playing commands such as volume, brightness and other media processing controls.

Quality of Service (QoS) Policy

Quality of service (QoS) policy is a set of rules that can be used to assign priority and other quality related options to different users or media flows, and guarantee certain levels of performance (such as bandwidth). QoS policy can be applied to media streamers (source devices) and media players (sink devices). The QoS policy can separate classes of traffic by assigning different services, such as streaming or data transfer, with traffic identifiers (TID). Traffic specifications (TSPEC) can be created, which contain a set of parameters that describe the desired characteristics of the traffic stream (TS).

Remote User Interface (RUI)

A remote user interface is a device within a network (address and processing function) that can send and receive device and service commands. UPnP control points can be used to control media device functions such as play, pause, stop, record and other control functions.

Chapter 7

Digital Rights Management (DRM)

Home media networks can use digital rights management (DRM) to control and provide copy protection to exert control over the distribution of digital media within a home or group of related people or devices. DRM involves the control of physical access to information, identity validation (authentication), service authorization (certificates), and media protection (encryption).

Digital rights management for home media networks involves the use of a content protection chain. The content protection chain links the rights that are included with media sources (such as TV broadcast signals or stored media), through distribution systems (set top box to digital video recorder), to viewing devices (such as television sets). A digital transmission content protection system (DTCP) can be used by home media networks to distribute and protect content.

Figure 7.1 shows that content protection in home is an extension of the content protection used in other sources, including broadcast and stored media systems. The content from TV broadcasters is protected by conditional access (CA) systems through a set top box (STB). Content from stored media (DVDs, Blu-Ray) is protected by a content scrambling system. Content that is distributed through the home is protected by digital transmission content protection (DTCP) and high definition content protection (HDCP). Content that is stored by users is protected by content protection for recordable media (CPRM).

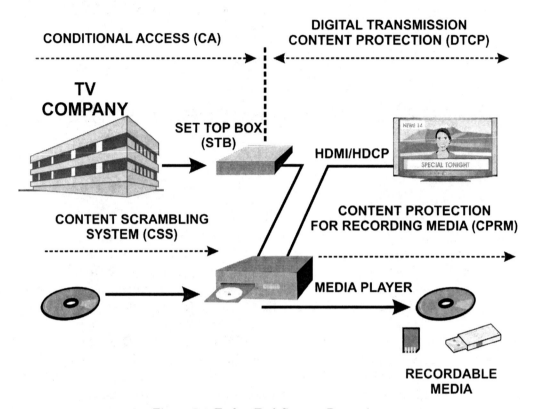

Figure 7.1, End to End Content Protection

Media Portability

Media portability is the ability to transfer media from one device to another. Media portability can range from storing locally received media in a hard disk, such as a personal video recorder, to sharing media through home connections, such as a premises distribution network.

Home media networks may be used by people to transfer media from home network devices (such as media servers) to portable devices (such as tablet PCs or smartphones). When users request to transfer content to portable devices, the user rights should be checked and included with the media as part of the transfer.

Chapter 7

Authentication

Authentication is the process of exchanging information between a communications device (typically a user device such as a mobile phone or computing device), and a communications network that allows the carrier or network operator to confirm the true identity of the user (or device). This validation of the authenticity of the user or device allows a service provider to deny service to users that cannot be identified. Thus, authentication inhibits fraudulent use of a device that does not contain the proper identification information.

Authentication credentials are the information elements that are used to identify and validate the identity of a person, company or device. Authentication credentials may include identification codes, service access codes and secret keys.

A common way to create identification codes that cannot be decoded is through the use of password hashing. Password hashing is a computational process that converts a password or information element into a fixed length code. Password hashing is a one way encryption process, as it is not possible to derive the original password from the hashed code.

Figure 7.2 shows how hashing security can be used to create identification codes that cannot be directly converted back into the original form. In this example, a hash code is created from the password 542678. The hashing process involves adding the odd digits to calculate a result (14), adding the even digits to calculate a result (18), and then storing the results of the password (1418). By using this hashing process, the original password cannot be recreated from the hashing results, even if the hashing process is known.

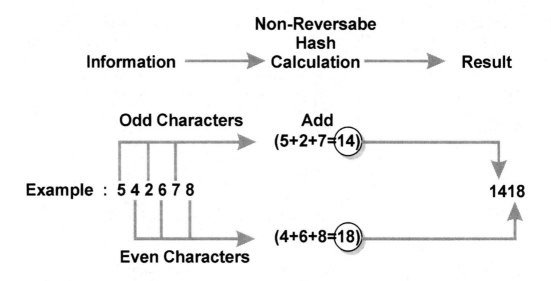

Figure 7.2, Password Hashing

Authentication can be performed without the need to transfer secret information over a communication channel. The authentication process only transfers the results of a calculation using the same secret information that is stored in each device (keys). The results of the calculation (Res) are transferred to the authenticator device which uses the same stored information (keys) in both devices to calculate its own result. If the calculated results match, this proves the authenticator has been validated. To ensure the process cannot be duplicated, an additional number is used each time an authentication is performed (a random number). That additional number is transferred from the authenticator to the device being authenticated.

Figure 7.3 shows a typical authentication process used in a DRM system. In this diagram, a DRM server wants to validate the identity of a user. The DRM system has previously sent a secret key to the user. The authentica-

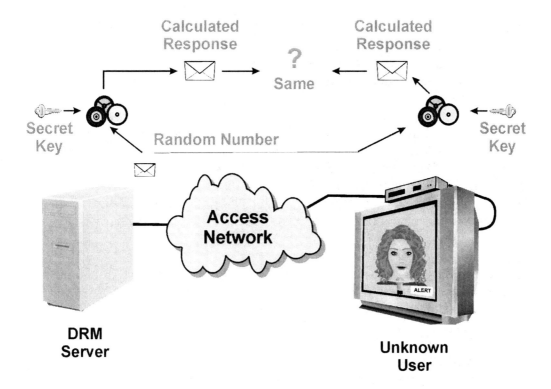

Figure 7.3, Authentication Operation

tion process begins with the DRM server sending an authentication request and a random number. This random number is used by the receiving device, and is processed with the secret key with an authentication algorithm (data processing) to produce a calculated result. This result is sent to the originator (authenticator). The originator uses the random number it sent, along with its secret key, to calculate a result. If the result received from the remote device matches its own result, the authentication passes. Note that the secret key is not sent through the communication network, and that the result will change each time the random number changes.

Digital Certificate

A digital certificate is information that is encapsulated in a file or media stream that identifies that a specific person or device has originated the media. Certificates are usually created or validated by a trusted third party that guarantees or assures that the information contained within the certificate is valid.

A trusted third party is a person or company that is recognized by two (or more) parties to a transaction (such as an online) as a credible or reliable entity who will ensure a transaction or process is performed as both parties have agreed. A trusted third party that issue digital certificates is called a certificate authority (CA). The CA typically requires the exchange of specific types of information between the parties to validate their identities before a certificate is issued.

The CA maintains records of the certificates that it has issued in repositories and these records allow the real time validation of certificates. If the certificate information is compromised, the certificate can be revoked.

Figure 7.4 shows how digital certificates can be used to validate the identity of a provider of content. This diagram shows that users of digital certificates have a common trusted bond with a certificate authority (CA). This diagram shows that because the content owner and content user both exchange identification information with the CA, they have an implied trusted relationship with each other. The content user registers with the CA and receives a certificate from the CA. The content owner registers with the CA and receives a key pair and a certificate signed by the CA. When the user requests information from a content owner, the content owner sends the public key that is in the signed certificate. Because the user can validate the signature on the certificate using the CA's public key, the user can trust the certificate, and use the public key provided by the content owner (such as an online store).

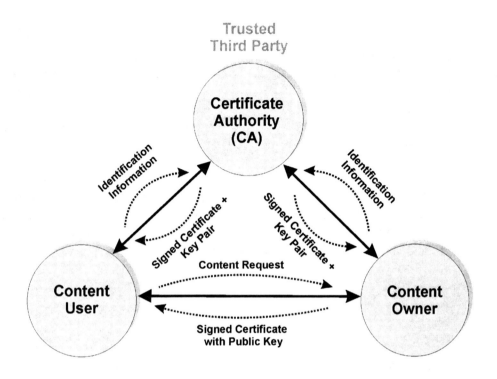

Figure 7.4, Digital Certificate Operation

Digital Signature

A digital signature is a number that is calculated from the contents of a file or message using a private key appended or embedded within the file or message. The inclusion of a digital signature allows a recipient to check the validity of the file or data by decoding the signature to verify the identity of the sender.

To create a digital signature, the media file is processed using a certificate, or form of validated identifying information, using a known encoding (encryption) process. This produces a unique key that could only have been created using the original media file and identifying certificate. The media file and the signature are sent to the recipient who separates the signature from the media file and decodes the key using the known decoding (decryption) process.

Encryption

Encryption is a process of a protecting voice or data information from being used or interpreted by unauthorized recipients. Encryption involves the use of a data processing algorithm (formula program) that uses one or more secret keys that both the sender and receiver of the information use to encrypt and decrypt the information. Without the encryption algorithm and key(s), unauthorized listeners cannot decode the message.

Figure 7.5 shows the basic process used by encryption to modify data to an unrecognizable form. In this example, the letters in the original information are shifted up by 1 letter (example - the letter I becomes the letter J). With this simple encryption, this example shows that the original information becomes unrecognizable to the typical viewer.

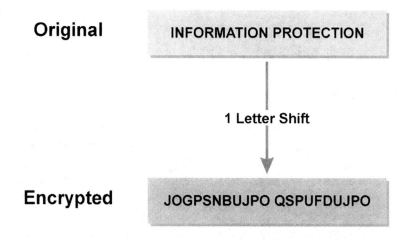

Figure 7.5, Basic Encryption Process

Encryption Keys

An encryption system typically uses a combination of a key (or keys) and encryption process (algorithm) to modify the data. An encryption key is a unique code that is used to modify (encrypt) data to protect it from unauthorized access. An encryption key is generally kept private or secret from other users. Encryption systems may use the same encryption key to encrypt and decrypt information (symmetrical encryption) or they may use different keys to encrypt and decrypt information (asymmetrical encryption). Information that has not been encrypted is called cleartext and information that has been encrypted is called ciphertext.

The encryption key length is the number of digits or information elements (such as digital bits) that are used in an encryption (data privacy protection) process. Generally, the longer the key length, the stronger the encryption protection will be.

Figure 7.6 shows how encryption can convert non-secure information (cleartext) into a format (cyphertext) that is difficult or impossible for a recipient to understand without the proper decoding keys. In this example, data is

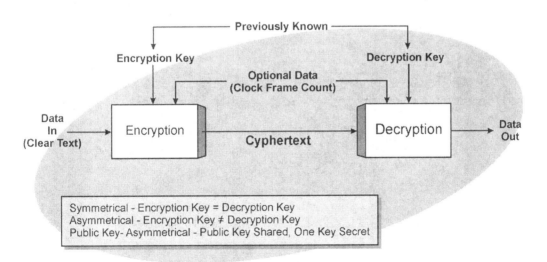

Figure 7.6, Encryption Operation

provided to an encryption processing assembly that modifies the data signal using an encryption key. This diagram also shows that additional (optional) information, such as a frame count or random number, may be used along with the encryption key to provide better information encryption protection.

Symmetric and Asymmetric Encryption

Encryption systems may use the same key for encryption and decryption (symmetric encryption) or different keys (asymmetric encryption). Generally, asymmetric encryption requires more data processing than symmetric encryption.

Figure 7.7 shows the differences between symmetric and asymmetric encryption. This diagram shows that symmetrical encryption uses the same keys to encrypt and decrypt data, and that asymmetric encryption uses different keys and processes to encrypt and decrypt the data.

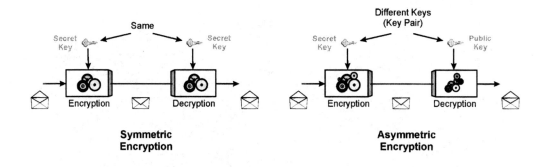

Figure 7.7, Symmetric and Asymmetric Encryption Processes

Public Key Encryption

The encryption process may be private or public. Public key encryption is an asymmetric authentication and encryption process that uses two keys, a public key and a private key, to setup and perform encryption between communication devices. The public key can be combined with the private key to increase the key length, providing a more secure encryption system. The public key is a cryptographic key that is used to decode information that was encrypted with its associated private key. The public key can be made available to other people and the owner of the key pair only uses the private key.

The encryption process may be continuous or it may be based on specific sections or blocks of data. A media block is a portion of a media file or stream that has specific rules or processes (encryption) applied to it. The use of blocked encryption may make re-synchronization easier, which may be necessary in the event of transmission errors.

Usage Restrictions

Usage restrictions are rules that limit how media or programs may be used. Usage restrictions may be included in metadata descriptions, or media may be encoded or wrapped in a way that only a media player with the correct keys can decode.

Figure 7.8 shows how digital rights management (DRM) may allow for media to be transferred from one device to another. In this example, a digital movie is downloaded to an IP Set Top Box (IP STB). The IP STB can then transfer the movie to a portable media player. The user will be allowed to transfer and view the movie, as long as the DRM information is set to allow the transfer and viewing, and the keys and viewing authorization information have not expired.

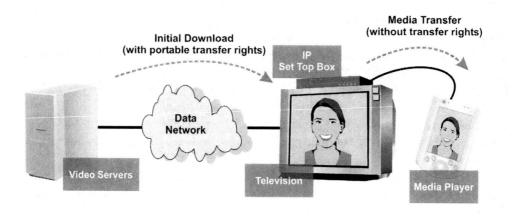

Figure 7.8, Media Portability

Copy Control Information (CCI)

CCI is data that represents the authorization and processes that may be used to copy content. CCI may be embedded into the content so it can be read and used by devices that may be used to copy content (such as a DVD player/recorder). CCI sends some data along with the media signal (copy control bits) that indicate whether or not the media can be copied.

CCI may be embedded (encoded) within the protected media stream to ensure that it cannot be altered. An encryption mode indicator (EMI) identifies how the CCI data is protected (unencrypted or encrypted).

Digital Transmission Content Protection (DTCP)

Digital transmission content protection is a set of processes that can be used to manage the rights of video content that are distributed through home networks. The DTCP system uses a combination of protocols, technical requirements and licensing processes to manage and enforce content protection.

DLNA commands define how content may be transferred and used between media devices that protect users from unauthorized copying, intercepting and tampering with media. The digital transmission licensing administer (DTLA) oversees the establishment and coordination of DTCP device certification and issuing of licenses.

The DTCP system can perform authentication, content protection (encryption), and system renewal capability. Access to the keys that are needed to decode the decryption processes and content are controlled by license certificates. Information about DTCP can be found at www.DTCP.com.

The content that is protected by a DTCP system may come from different sources including TV broadcast content and stored media. The conditional access (CA) system protects content on broadcast distribution systems (such as cable TV or DTT systems). Media that is provided from stored devices (such as DVD or Blu-ray) is protected by a content protection for recordable media (CPRM) system. The DTCP content distribution system is used from the access device (such as a set top box, media server, or Blu-ray player) to the end user device (such as a monitor or television).

Authentication and Key Exchange (AKE)

Devices within a DTCP enabled system perform authentication to validate their identity and authorization to receive and process media (using the device certificate). The DTCP system may perform full authentication or restricted authentication. Full authentication is a secure process that can be used for all content and processes. Restricted authentication is a reduced security process (used on devices with limited authentication processing capabilities) that can be used on content that is indicated as "copy-once" or "no more copies".

Device Certificates

Each DTCP device contains a unique device certificate. When DTCP devices are connected to each other, the device certificates are exchanged with each other to setup a secure (protected) communication channel.

Content Protection (Encryption)

Once the identity and authorization of devices are validated, content encryption keys can be exchanged. Encryption modifies content using data scrambling processes (cryptographic functions) and security keys. The DTCP system encrypts content to create a secure communication channel that can be used to transfer protected content between devices. The encryption process on the protected content channel that is transferred between the devices is continually changed (keys are constantly updated). The content rights information (CCI data) can be included in the protected content channel.

System Renewability

The DTCP system was designed with the ability to alter its security information (certificates) and processes over time. This includes key renewability and key revocation. Key renewability is the ability of an encryption system to issue new keys that can be used in the encoding or decoding of information. Key revocation is the process of deleting or modifying a key so a user or device is no longer able to decode content. Key revocation may be performed by sending a command from a system to a device. The DTCP system processes (security protocols) may also be updated as new security methods are discovered and distributed.

Figure 7.9 shows how a digital transmission content protection system can be used to protect content that is transferred between devices that are connected by a home media network. This example shows that the source device (home media server) has been requested to send content to a sink device (a television set). To enable the connection, authentication of the television set is performed to validate its identity and authorization to receive protected content (using the device certificate). After the television set has been validated, content encryption keys are exchanged, allowing the media server to encrypt the content and the television to decrypt and display the content. This example shows that the keys that protect the content are periodically changed over time.

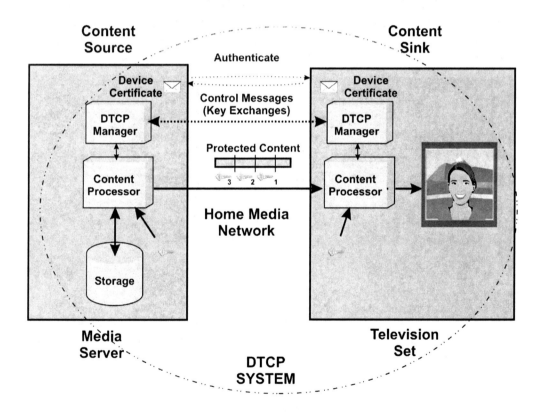

Figure 7.9, Digital Transmission Content Protection (DTCP) System

Digital Transmission Control Protection Plus (DTCP+)

Digital transmission control protection plus is an evolution of the DTCP system that was released in November 2010. The DTCP+ system includes the use of a digital only token, media independent (agnostic) content transport process, content management information (CMI) process, copy control information (CCI), and remote access control.

High Bandwidth Digital Content Protection (HDCP)

High bandwidth digital content protection is the commands and processes that are used by the digital video interface (DVI) system to perform authentication, encryption and authorization of services. The HDCP system allows one device to identify another and determine what type of device it is (a display or a recorder).

HDCP may be used with an HDMI connection. HDMI is a set of commands and processes (protocols) that can be used to transfer uncompressed (high speed digital) media between devices. Not every device with an HDMI input conforms to the HDCP industry standards (specifications).

There are multiple versions of the HDCP specification. The HDCP specification has evolved to allow more inputs to a viewing device and to allow for repeaters that can extend the distance between a digital content source (such as a STB) and a digital content player (such as a television).

The HDCP process begins when a user or device requests to play or use media. The requesting device (a TV monitor being plugged into a Blu-ray player) starts to authenticate (validate) its identity with the sender (source) of the signal (the Blu-ray player). This process involves the use of random numbers and cryptographic processes that should produce the same result on the receiving device (the TV) and the sending device (the Blu-ray player). This process is sometimes called a handshake.

If the authentication process is successful, key codes can be exchanged between the devices that allow the sender to encode the media and the receiver to decode the media. The keys are only valid for short periods of time, and new keys must be regularly created and exchanged.

Appendix 1 - Acronyms

AP-Access Point
A/V-Audio Visual
ABLP-Adaptive Bit Loading Protocol
ACPI-Advanced Configuration and Power Interface
ADTDM-Adaptive Time Division Multiplexing
AES-Advanced Encryption Standard
AKA-Authentication and Key Agreement
AKE-Authentication Key Exchange
App Stores-Application Stors
ATM-Asynchronous Transfer Mode
AVT-AV Transport Service
BAV-Base Audio Visual
BER-Bit Error Rate
BitTorrent-Bit Torrent
BL-ECM-Baseline Entitlement Control Messages
BPL-Broadband Over Powerline
CA-Central Arbiter
CA-Conditional Access
CAT-Cable Category
CB-Citizens Band
CCI-Copy Control Information
CDS-Content Directory Service
CEBus-Consumer Electronics Bus
CEPCA-Consumer Electronics Powerline Communication Alliance
Ciphertext-Cipher Text
CM-Connection Manager Service
Coax Amp-Coaxial Amplifier
CP-Content Protection
CP-Control Point
CP-Copy Protection
CPMS-Copy Protection Management System
CPRM-Content Protection For Recordable Media
CRL-Certificate Revocation List
CSMA-Carrier Sense Multiple Access
CSMA/CA-Carrier Sense Multiple Access/Collision Avoidance
CWDM-Coarse Wave Division Multiplexing
DAP-Domain Access Point
DCM-Device Control Module
DES-Data Encryption Standard
DHN-Digital Home Network
DHS-Digital Home Standard
DLNA-Digital Living Network Alliance
DM-Domain Master
DMA-Digital Media Adapter
DMC-Digital Media Controller
DMP-Digital Media Player
DMPr-Digital Media Printer
DMR-Digital Media Renderer
DMS-Digital Media Server

DRM-Digital Rights Management
Drop Amp-Drop Amplifier
DTCP-Digital Transmission Content Protection
DTCP+-Digital Transmission Control Protection Plus
DTLA-Digital Transmission Licensing Administrator
DTT-Digital Terrestrial Television
ECCA-Enhanced Distributed Channel Access
EDCA-Enhanced Distributed Channel Access
EMI-Encryption Mode Indicator
EP-End Point
EPF-Electronic Picture Frame
ESA-External Scrambling Algorithm
ESVP-Extended Secure Video Processing
FBN-Frequency Band Notch
FDQAM-Frequency Diverse Quadrature Amplitude Modulation
FMC-Fixed Mobile Convergence
FPA-Front Panel Assembly
GM-Global Master
Gross Rate-Gross Data Rate
GW-Gateway
HAVi-Home Audio Video Interoperability
HCCA-HCF Coordination Channel Access
HCCA-Hybrid Coordination Function Controlled Channel Access
HCNA-HPNA Coax Network Adapter
HD-High Definition
HDCP-High Bandwidth Digital Content Protection
HDD-Hard Disk Drive
HDMI-High Definition Multimedia Interface
HD-PLC-High Definition Power Line Communication
HDTV-High Definition Television
HID-Home Infrastructure Device
HMN-Home Media Network
HND-Home Network Device
HomePlug Access BPL-HomePlug Access Broadband Power Line
HomePlug AV-HomePlug Audio Visual
HomePNA-Home Phoneline Networking Alliance
HPCC-HomePlug Command & Control
HSIA-High Speed Internet Access
HTTP-Hypertext Transfer Protocol
HVN-Home Video Network
IAV-Intermediate Audio Visual
IH-In Home
IHDN-In-Home Digital Networks
Inline Amp-Inline Amplifier
InstanceID-Instance Identifier
IP Audio-Internet Protocol Audio
IP Video-Internet Protocol Video
IPTV-Internet Protocol Television
IR Routing-Infrared Routing
ISMA-Internet Media Streaming Alliance
ISP-Inter-System Protocol
LAN-Local Area Network
LASER-Light Amplification By Stimulated Emission of Radiation
LED-Light Emitting Diode

LLC-Logical Link Control
LLTD-Link Layer Topology Discovery
MAC-Media Access Control
MAC Protocol-Medium Access Control Protocol
MAP-Media Access Plan
MHD-Mobile Handheld Device
MIMO-Multiple Input Multiple Output
MIU-Media Interoperability Unit
MMF-Multimode Fiber
MoCA-Multimedia over Coax Alliance
MoDem-Modulator/Demodulator
MOST-Media Oriented Systems Transport
MP4-MPEG-4
MRD-Marketing Requirements Document
MS-Media Server
NAT Traversal-Network Address Translation Traversal
NCF-Network Connectivity Function
Net-Internet
Net Rate-Net Data Rate
Network ID-Network Identifier
NNW-No New Wires
NSA-Native Scrambling Algorithm
NTD-Network Topology Discovery
NUT-Net UDP Throughput
OFDM-Optical Frequency Division Multiplexing
OPERA-Open PLC European Research Alliance
P2P-Peer to Peer
Parametric QoS-Parametric Quality of Serivce
PDN-Premises Distribution Network
PER-Packet Error Rate
PKE-Public Key Encryption
PLC-Power Level Control
PLC-Power Line Carrier
PLC-Powerline Communications
POE-Point of Entry
PON-Passive Optical Network
Priority Based QoS-Priority Based Quality of Service
PSD-Power Spectral Density
PSD Mask-Power Spectral Density Mask
QC-LDPC-Quasi-Cyclic Low-Density Parity Check
QoE-Quality of Experience
QoS-Quality Of Service
QoS Policy-Quality of Service Policy
RAND-Reasonable and Non-Discriminatory
RCS-Rendering Control Service
RFIC-Radio Frequency Integrated Circuit
RTP-Real Time Transport Protocol
RUI-Remote User Interface
S-Slave
SAC-Secure Authenticated Channel
SD-Service Discovery
SD-Standard Definition
SMF-Single Mode Fiber
SNAP-Sub Network Access Protocol
SRM-System Renewability Messages
SVP-Secure Video Processing

SVPLA-Secure Video Processor Licensing Authority
THIG-Topology Hiding Inter-network Gateway
TOS-Terms of Service
TOS-Type Of Service
TRS-Tamper Resistant Software
TXOP-Transmission Opportunity
UNIT-Unified Network Interface Technology
UPA-Universal Powerline Association
UPnP-Universal Plug and Play
UPnP AV-Universal Plug and Play Audio Visual
USB-Universal Serial Bus
USM-User Services Management
Web Access API-Web Access Application Programming Interface
Wi-Fi-Wireless Fidelity
Wi-Fi TV-Wi-Fi Television
WLAN-Wireless Local Area Network
WMM-Wi-Fi Multimedia

Index

802.11a, 43
802.11e, 74-75
802.11g, 8, 26, 43
802.11n, 8, 26, 43, 74-75
Access Point (AP), 41, 44
Access Protocol, 89, 91
Adaptive Modulation, 29-30
Advanced Encryption Standard (AES), 59
Alien Networks, 2, 9, 15, 77-78
Asymmetric Encryption, 102
Authentication Key Exchange (AKE), 105
Automatic Addressing, 89, 91
Average Data Rate, 3, 18
Bit Error Rate (BER), 4, 19
Broadband Over Powerline (BPL), 62, 81
Capabilities Description, 89
Carrier Sense Multiple Access (CSMA), 58-59
Central Arbiter (CA), 58-59, 75, 93, 98, 105
Channel Bonding, 29, 35-36, 42, 75
Channel Multiplexing, 35
Channel Selection, 34
Cipher Text (Ciphertext), 101
Citizens Band (CB), 57
Cleartext, 101
Co-Existence, 1, 3, 9, 13, 16, 48, 62, 77, 81
Conditional Access (CA), 58-59, 93, 98, 105
Configuration, 5-9, 21, 23-26, 61, 89
Connectivity Layer, 87
Consumer Electronics Bus (CEBus), 45, 61
Content Protection (CP), 5-6, 9, 22-23, 93-94, 105-108
Content Protection For Recordable Media (CPRM), 93, 105
Content Rights, 6, 22, 106
Contention Based, 33
Contention Free, 33, 67
Control Point (CP), 91-92
Copy Control Information (CCI), 104, 106, 108
Copy Protection (CP), 93
Cross Phase Coupling, 59
Data Encryption Standard (DES), 56, 70
Data Throughput, 4, 9, 19
Data Transfer Rate, 6, 8
Demultiplexing, 35
Device Certificate, 105-107
Device Discovery, 86-87, 89

115

Digital Home Standard (DHS), 55, 62-63
Digital Living Network Alliance (DLNA), 83-88, 105
Digital Media Controller (DMC), 84-85
Digital Media Player (DMP), 84-85
Digital Media Printer (DMPr), 85
Digital Media Renderer (DMR), 84-85
Digital Media Server (DMS), 84-85
Digital Only Token, 108
Digital Rights Management (DRM), 93, 96-97, 103
Digital Terrestrial Television (DTT), 105
Digital Transmission Content Protection (DTCP), 93, 105-108
Digital Transmission Control Protection Plus (DTCP+), 93, 105-108
Digital Transmission Licensing Administrator (DTLA), 105
Domain, 77-78, 89
Domain Master (DM), 77-78
Double Hop, 11
Echo Cancellation, 31
Echo Control, 31
Electrical Noise, 45
Encryption, 6, 22-23, 56, 59, 63, 70, 93, 95, 99-108
Encryption Mode Indicator (EMI), 104
Enhanced Distributed Channel Access (EDCA), 75

Error Rate, 4, 19
Ethernet, 1, 5, 14, 20, 39-40, 42, 61, 64, 83, 87
Event Notifications, 90-91
Fiber Type, 53
Firewall, 7, 25
FireWire, 64
Foreign Networks, 77
Frequency Band, 1, 14, 33, 50, 57, 68, 73
Frequency Band Notch (FBN), 1, 14
Frequency Diverse Quadrature Amplitude Modulation (FDQAM), 66
Gateway (GW), 3, 8, 12, 7, 24-25, 32, 65, 67
Global Master (GM), 77-78
Gross Data Rate (Gross Rate), 3, 9, 18
Hashing, 95-96
HCF Coordination Channel Access (HCCA), 75
High Bandwidth Digital Content Protection (HDCP), 93, 108
High Definition (HD), 7, 11, 63, 93
High Definition Multimedia Interface (HDMI), 108
High Definition Power Line Communication (HD-PLC), 63
High Definition Television (HDTV), 3, 7
Home Coverage, 5, 9, 21
Home Media Management, 83
Home Media Streaming, 11-12
Home Optical Networks, 53

Index

Home Phoneline Networking Alliance (HomePNA), 2, 15, 55, 64-70
Home Television Service, 7, 9
Home Theater, 11
HomeGrid, 76-78
HomePlug 1.0, 55-59, 61
HomePlug Audio Visual (HomePlug AV), 58-61
HomePlug Specification, 55, 61
HomePNA 1.0, 65
HomePNA 2.0, 2, 15, 65-67
HomePNA 3.0, 66-68
HomePNA 3.1, 2, 15, 67-69
Hybrid Coordination Function Controlled Channel Access (HCCA), 75
Hypertext Transfer Protocol (HTTP), 12, 83, 86
In Home (IH), 2, 15, 93
Incomplete Visibility, 5, 21
Installation Costs, 8-9, 26
Installation Skills, 6-7, 9, 23-24
Inter-System Protocol (ISP), 77
Interference Avoidance, 29, 33
Interference Detection, 34
Interfering Signals, 49, 51, 65
Internet Access, 3, 5-6
Internet Protocol Television (IPTV), 8, 5, 22, 40, 58, 73, 81
Intersystem Interference, 1, 9, 14
Jitter, 1, 8, 10-11, 4-5, 9, 19-20, 32, 59
License Costs, 8-9, 26
Light Amplification By Stimulated Emission of Radiation (LASER), 53, 80, 85
Light Emitting Diode (LED), 53, 79-80
Link Detection, 37
Local Area Network (LAN), 1-2, 13, 15, 39-40, 42-44, 72, 74-75
Management Layer, 86-87
Master Control, 50, 67, 92
Master Controller, 32
Master Slave Relationship, 32
Media Access Control (MAC), 59, 75, 91
Media Access Plan (MAP), 33, 67
Media Cycle, 67
Media Device, 5, 21, 92
Media Format Layer, 86-87
Media Management Layer, 86-87
Media Oriented Systems Transport (MOST), 1, 14, 42, 56, 61, 71
Media Portability, 94, 104
Media Renderer, 84-85, 92
Media Server (MS), 12, 84-85, 91, 105, 107
Media Transport Layer, 86-87
Modulation Type, 29-30, 66
Modulator/Demodulator (MoDem), 1, 8, 7, 24, 40-41, 58, 60, 67
MPEG-2, 7, 3-4, 18-19, 86
MPEG-4 (MP4), 7, 3, 18
Multimedia over Coax Alliance (MoCA), 55, 70-72, 81
Multiple Input Multiple Output (MIMO), 74-76
Net Data Rate (Net Rate), 3, 9, 18

117

Network Coordination, 34
Network Mapping, 37
Network Stack Layer, 87
Network Topology Discovery (NTD), 36
No New Wires (NNW), 2-3, 9, 16-17, 72
Open PLC European Research Alliance (OPERA), 62, 81
Optical Frequency Division Multiplexing (OFDM), 57, 70
Packet Analysis, 7, 25
Packet Error Rate (PER), 4-8, 10, 4, 19, 30, 66
Peak Data Rate, 4, 19
Performance Monitoring, 73
Phase Coupling, 45, 59
Point of Entry (POE), 1, 14
Power Level Control (PLC), 29, 34-35, 45, 52, 62
Power Line Carrier (PLC), 45, 62
Powerline Communications (PLC), 45, 62
Premises Distribution Network (PDN), 94
Priority Levels, 75
Privacy, 5, 9, 22, 63, 70, 101
Protocol Adaptation, 7, 9, 24-25
Public Key Encryption (PKE), 103
Quality Of Service (QoS), 1, 4-5, 9, 13, 19-20, 42, 60, 66-67, 70, 73-74, 76, 91-92
Quality of Service Policy (QoS Policy), 92
Real Time Transport Protocol (RTP), 1, 14, 83
Remediation, 7, 9, 24
Remote Diagnostics, 7, 9, 24-25
Remote User Interface (RUI), 92
Rendering Control, 91-92
Rights Management, 5-6, 9, 22-23, 93, 103
Self Configuration, 6, 23
Self Discovery, 6, 23
Service Discovery (SD), 89, 91
Signal Leakage, 1, 14, 35, 51-52
Signal Level Detection, 34
Signal Reflection, 31, 56
Signature, 98-99
Single Mode, 53, 79-80
Single Mode Fiber (SMF), 79-80
Site Review, 2-3, 9, 16-17
Slave (S), 1, 8, 14, 26, 32, 42, 67, 92, 98, 100
Smart Antenna, 43-44, 74
Splitter Jumping, 71
Standard Definition (SD), 8
Streaming, 3, 5, 8, 11-12, 7, 25, 35, 62, 91-92
Support Costs, 8-9, 26-27
Symmetric Encryption, 102
Synchronized Transmission, 29, 32-33, 67
System Renewability, 106
Terms of Service (TOS), 32
Token, 40, 108
Topology, 2, 16, 36-37, 45-46
Topology Discovery, 36-37

Index

Transmission Delay, 4, 9, 19-20
Transport Layer, 86-87
Truck Roll, 8, 26
Type Of Service (TOS), 32
Unified Network Interface Technology (UNIT), 32-33, 67
Universal Plug and Play (UPnP), 83, 86, 88-92
Universal Plug and Play Audio Visual (UPnP AV), 83, 86, 91
Universal Powerline Association (UPA), 62, 81
Universal Serial Bus (USB), 61, 64, 90
Upgradability, 7, 24
Usage Restrictions, 103
Video Streaming, 5, 35
Visibility, 5, 9, 21
Wi-Fi Television (Wi-Fi TV), 41, 44
Wired LAN, 39-40
Wireless Fidelity (Wi-Fi), 40-41, 44, 72-75
Wireless Local Area Network (WLAN), 3, 18, 40-44, 55, 73-75, 83
X-10, 45-47, 59, 61

CPSIA information can be obtained at www.ICGtesting.com
Printed in the USA
LVOW111639050812

293000LV00012BA/235/P